TEST YOUR HIGHER MATHEMATICS

by

P.W. Westwood,
Principal Teacher of Mathematics,
Kirkcaldy High School.

and

C.M. Stewart

ISBN 0 7169 3247 4

ROBERT GIBSON · Publisher
17 Fitzroy Place, Glasgow, G3 7SF, Scotland, U.K.

HOW TO USE THIS BOOK

THE PURPOSE OF THE TEXT

This book is designed to reinforce the basic knowledge and skills on which problem solving in Higher Mathematics is based, by providing systematic revision of each unit (1, 2 and 3) and of the whole course. A little revision done regularly and continuously is known to be far more effective than a massive last-minute effort.

RATE OF WORK

Teachers and individual pupils are welcome to use this book as they see best, but the intention of the authors is to provide 36 sets of examples for each unit, one set per night, four nights per week, for nine weeks. For each unit, there are approximately eight sets on revision of previous work, six sets on each topic, and four miscellaneous sets on the whole unit.

REVISION

Revision of previous work at the start of each unit is included for two main reasons. Firstly, it is essential that pupils are confident with their prior knowledge (especially with Standard Grade work at the start of Unit 1). Secondly, we see merit in using the book continuously throughout the course (with a minimal break between successive units).

CHANGING THE ORDER

The order in which the four topics within each unit are taken may vary, so it may be necessary to attempt the sets in a different order from that in which they are printed. Sensible reference to the contents list should avoid any problem in this respect. Each topic should be completed in class prior to attempting the revision set on it, as each set on a topic will generally cover the whole of that topic.

CORRECTION OF ANSWERS

The user should correct each set immediately upon completion and record the score. If a whole class works through the sets together, it is beneficial to go over each evening's work the next day. Individual pupils working through the book should ask their teacher or tutor about any questions they cannot solve themselves.

DIFFERENTIATION OF WORK

Each set consists of eight questions. An attempt has been made to keep questions 1 to 5 in each set at level C and questions 6 to 8 at the A / B level. This has, however, not always been possible, as some topics only occur at the one level, but the structure used does allow the work to be reduced (if so desired) for those aiming only at a C pass.

USE OF CALCULATORS

The final exam has a paper where use of a calculator is permissible and another paper where it is not. Exam questions are therefore of three possible types; those where the calculator is prohibited, those where it is necessary and those where it makes no difference. For simplicity, this text identifies only those examples where the calculator is required and all other examples should be treated as possibly occurring in Higher Paper I (the non-calculator paper). You should, therefore, use your calculator **only** where you see the calculator icon displayed to the left of the question.

WHOLE COURSE REVISION

Note that Sets 109–120 focus on facts and formulae and can be done much more quickly than the preceding sets, as the authors would expect candidates to be devoting most of their time at this stage to working through previous Higher Mathematics examination papers.

CONTENTS

UNIT 1

REVISION OF STANDARD GRADE ESSENTIALS

SET 1

1. Simplify $3a^{1/2} \times 3a^{-1/2}$
2. Expand $(x + 1)(x - 2)(x + 4)$
3. Evaluate $(2\sqrt{3} + \sqrt{7})(2\sqrt{3} - \sqrt{7})$
4. Write down the coordinates of the mid-point of AB where A is (–1, 2) and B is (3, 4).
5. The line $2x - 5y = a$ passes through the point (3, –1). Find a.
6. Solve $x^2 - 6x = 0$
7. Calculate the area of triangle PQR in which PQ = 4 cm, QR = 5 cm and $P\hat{Q}R = 50°$.
8. State the coordinates of a maximum turning point on the curve $y = 2 \cos x - 1$, where x is a real number.

SET 2

1. Solve the equations (i) $2^{(x^2)} = 16$ (ii) $3^{-x} = 2^5 - 5$, (x real).
2. Factorise fully $18x^2y - 33xy^2 - 30y^3$
3. Expand $\left(\sqrt{x} + \frac{1}{\sqrt{x}}\right)^2$
4. Write down the gradient of the line $5x + 7y = 12$.
5. The line $\frac{x}{4} + \frac{y}{5} = 1$ cuts the x-axis at P and the y-axis at Q. Calculate the area of triangle OPQ.
6. State the maximum value of the function $7 - x^2$.
7. Simplify $\frac{\sin (180 - x)°}{\sin (90 - x)°}$
8. In triangle ABC, AB = $\sqrt{2}$ cm, AC = $\sqrt{3}$ cm and BC = 3 cm. Calculate the value of cos A.

SET 3

1. Simplify $x^{1/2}(x^{1/2} + x^{-1/2})$
2. Factorise fully $x^4 - 5x^2 + 4$
3. Simplify (i) $\sqrt{75} - \sqrt{48}$ (ii) $\frac{\sqrt{27} + \sqrt{3}}{\sqrt{3}}$
4. Write down the equation of the line through the point (5, 6)
   (i) parallel to the x-axis (ii) parallel to the y-axis.

5. A line PQ makes an angle of 40° with the positive x-axis. Line RS is perpendicular to PQ. Give the size of the angle which RS makes with the positive x-axis.

6. List the solution set of $9 > x^2$ where x is an integer.

7. The sides of a triangle are 5, 6 and 7 units respectively. Calculate the size of the largest angle in the triangle.

8. For what value of x, $0 \leq x < 360$, does $5\sin(x - 40)°$ have its maximum value?

SET 4

1. Simplify (i) $\left(\frac{1}{625}\right)^{-1/4}$ (ii) $27^{2/3} \times 243^{-2/5}$

2. Factorise fully: (i) $2x^2 - 8xy$ (ii) $x^2 - 5x - 6$ (iii) $x^3 - x^2 - 2x$

3. Simplify $\dfrac{\sqrt{x} + \sqrt{x^3}}{\sqrt{x}}$, $x \neq 0$

4. Calculate the area bounded by the lines $y = 6$, $y = x$ and $y = -x$.

5. In ΔABC, $a = 5$, $b = 8$. State the range of possible values for c.

6. Solve the inequality $2(3x - 1) + 3 > 7 - 3(3x - 2)$.

7. In triangle ABC, $\sin \hat{A} = 0{\cdot}8$, $\sin \hat{B} = 0{\cdot}6$ and $a = 6$ cm. Calculate b.

8. The line $y = 10 - 2x$ cuts the x-axis at A and the y-axis at B. Find the coordinates of the mid-point of AB.

SET 5

1. Simplify $3x^2y^{-1/2} \, . \, (2xy^{-1/3})^{-2}$

2. Expand $(x + 3)(x - 1)(x + 2)$

3. Simplify $\dfrac{\sqrt{75} - \sqrt{50}}{\sqrt{12} - \sqrt{8}}$

4. Calculate the length of XY where X is the point (1, –1) and Y is (5, 3), leaving your answer as a surd.

5. State the general equation which describes
   (i) the set of lines with gradient 3
   (ii) the set of lines passing through (0, 5).

6. Solve the system of equations $2x + 3y = -5$, $5x - 2y = 16$.

7. Find the solution set of the inequality $\sin\theta > \frac{1}{2}$, where $0 \leq \theta \leq \pi$.

8. Express sin 402° as the sine of an acute angle and as the cosine of an acute angle.

SET 6

1. Evaluate $a^{2/3} \cdot b^{-1/4}$ when $a = 27$ and $b = 16$.
2. Simplify $(3x + 4y)^2 - x(x - 2y)$
3. Simplify $\frac{15}{\sqrt{3}} - \sqrt{27}$
4. Write down the coordinates of the mid-point of AB where A is (6, 4) and B is (–4, –6).
5. State the gradient of the line $\frac{x}{4} + \frac{y}{7} = 2$.
6. Find the solution set of the equation $x^2 - 5x = 36$, where x is real.
7. State the maximum and minimum values of $5 - \sin x$.
8. If $\tan X\hat{Y}Z = \frac{3}{4}$, calculate the area of ΔXYZ.

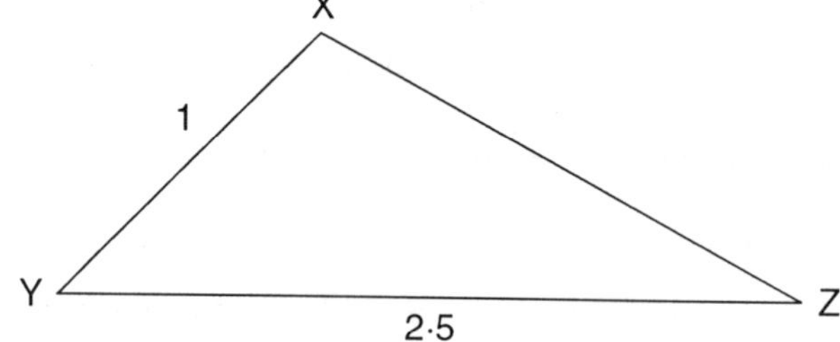

SET 7

1. Establish a formula for the shaded area.

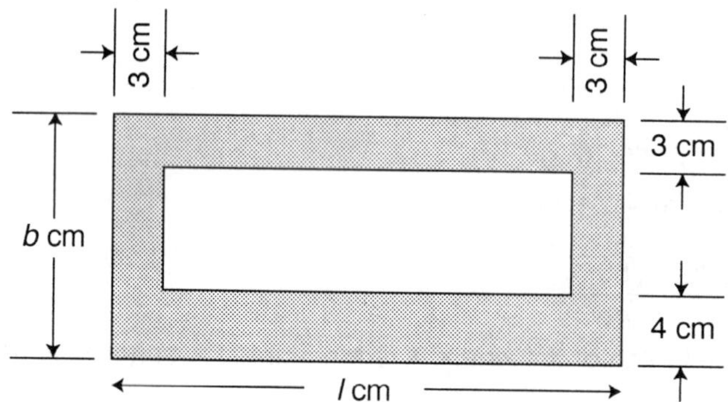

2. Expand $(a - b)(a + 2b - c)$.
3. State the maximum and minimum values of $\cos x$.
4. XY = 5, YZ = 3 and XZ = 8. What can be said about the points X, Y and Z?
5. Solve the system of equations $\left.\begin{aligned} 2x - 3y &= 3 \\ -4x + 2y &= 10 \end{aligned}\right\}$
6. The line $y = 4 - 2x$ cuts the x-axis at P and the y-axis at Q. Calculate the area of triangle OPQ.
7. Simplify $\frac{x^2 - 4}{6x - 9} \times \frac{2x^2 - 3x}{3x^2 + 5x - 2}$
8. Calculate the area of triangle ABC, given $a = 4$ mm, $b = 5$ mm and $\hat{C} = 30°$.

SET 8

1. If $3^x = \frac{1}{243}$, find x.
2. If $a = 2$, $b = -8$ and $c = -7$, calculate $\sqrt{b^2 - 4ac}$
3. Expand $(2x + 3)(x - 2)(3x + 5)$
4. Evaluate $c(c^2 - 5)$ when $c = 1 + \sqrt{2}$
5. Write down the exact value of sin 600°.
6. Solve the equation $x^4 - 8x = 0$, x a real number.
7. Simplify $\sin(90° - A) \, . \, \tan(360° - A) \, . \, \sin(180° - A)$
8. P is the point (3, 4). Write down the equation of the line through P
   (i) parallel to the x-axis
   (ii) parallel to the y-axis
   (iii) through the origin.

SET 9

1. Subtract $(3x^2 + 5x - 8)$ from $(4x^2 - 6x - 3)$.
2. Write down the exact value of cos 315°.
3. Simplify $\frac{2x^2 - 7x + 3}{4x - 2}$, $x \neq \frac{1}{2}$
   [Note: If $x = \frac{1}{2}$, then $4x - 2 = 0$ and the given fraction is not finite.]
4. Find x if $(x^{1/3} + 1)^{1/2} = 2$.
5. $V = \frac{\pi r^2}{3}(2r + 3h)$. Change the subject to h.

6. In the circle, radius 6 cm, $A\hat{O}B = 60°$. Calculate, correct to 1 decimal place,
   (i) the length of arc AB
   (ii) the area of sector AOB
   (iii) the area of ΔAOB.

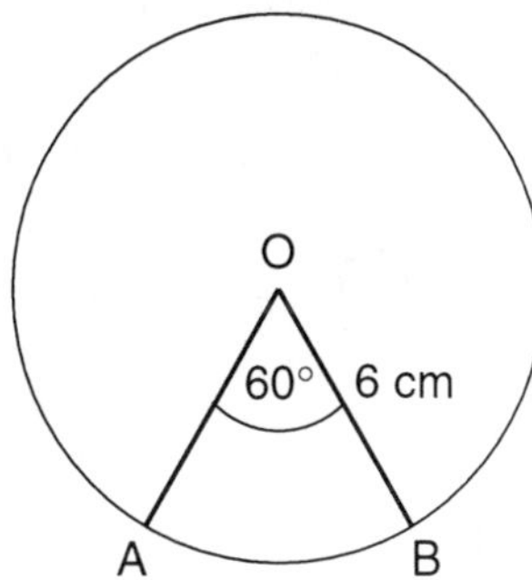

7. Sketch the graph of $1 - \sin x°$, $0 \leq x \leq 360$.
8. Find the points of intersection of the curve $y = 6 + x - x^2$ and the x-axis.

SET 10

1. Simplify $(x-3)^2-(x+3)^2$.
2. $(a, 7)$ is a point on the parabola $y = 16 - x^2$. Find a.
3. For what value of x does $17 \cos(x+25)°$ have its minimum value?
4. Factorise $(a-b)^2-(b-c)^2$.
5. Write down the coordinates of the mid-point of the line joining the points $(p+q, p-q)$ and $(p-q, q-p)$.
6. *(a)* Given $v = u + rt$, express r in terms of v, u and t.

   (b) Make F the subject of the formula $C = \frac{5}{9}(F-32)$.

   (c) Change the subject of the formula $v^2 = u^2 - 2gs$ to g.
7. A is (3, 1), B is (10, –3) and C is (11, 5). ABCD is a parallelogram. Find the coordinates of D.
8. Sketch the graphs of (i) $y = x^2$ (ii) $y = x^2 - 1$ (iii) $y = (x-1)^2$

SET 11

1. Simplify $(2^{1/3}-1)(2^{2/3}+2^{1/3}+1)$.
2. Evaluate $\frac{a^2}{1+a}$ when $a = \sqrt{3} - 1$.

3. Calculate the smallest angle in the triangle with sides 2, 3 and 4 cm.
4. Solve the equation $x(x-4) = x$, where x is a real number.
5. Sketch the graph of $y = \cos 2x° + 1$ for $0 \le x \le 360$.
6. The distance between the points (7, –14) and $(a, 26)$ is 41 units. Calculate the possible values of a.
7. P and Q are points $(2, a)$ and $(5, b)$ on the curve $xy = 20$. Calculate the gradient of PQ.
8. Simplify $\frac{a^n \times (a^{n-1})^n}{a^{n+1} \times a^{n-1}}$

SET 12

1. Solve the equation $x^2 - 9 = 2x + 1$.
2. Give the values of x for which each of the expressions

   (i) $\frac{1}{x^2-4}$ (ii) $\frac{1}{x^2-5x+6}$ is undefined.
3. Find the value of x for which $7 \cos(x+300)°$ has its maximum value.
4. A is the point $(t+u, u)$ and B is $(u-t, t)$. M is the mid-point of AB. Find the gradient of OM.
5. Find the exact solutions of $3 \sin x° = \sqrt{3} \cos x°$ for $0 \le x \le 360$.

The diagram shows the graph of $y = x(3 - x)$. Questions 6, 7 and 8 refer to the diagram.

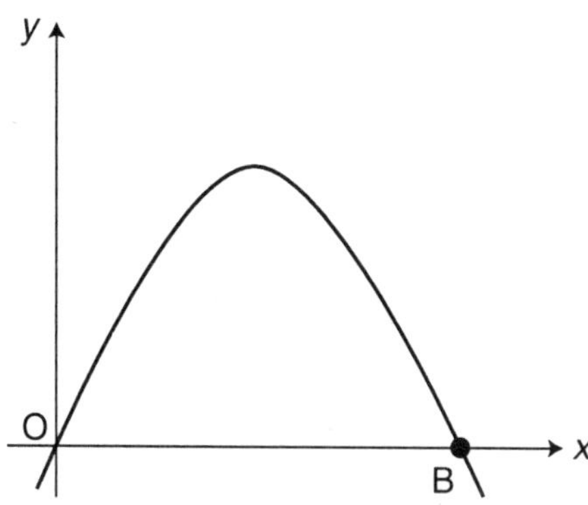

6. Give the coordinates of
   (i) B (ii) the turning point.

7. State the equation of the axis of symmetry.

8. Sketch the graphs of
   (i) $-f(x)$ (ii) $f(x) - 1$

THE STRAIGHT LINE

SET 13

1. Find p if $(5, p)$ lies on the line $2x + 6y - 7 = 0$.

2. A is (1, 2) and B(2, 6). Find the angle between AB and the x-axis.

3. Write in the form $ax + by + c = 0$ the equation of the straight line through $\left(0, -\frac{2}{3}\right)$ with gradient $\frac{4}{3}$.

4. A is (2, 1), B(3, 4), C(5, 7) and D(8, –2). P and Q are the mid-points of AC and BD respectively. Find the gradient of PQ.

5. Find the equation of the line through (3, –2) and (5, 1).

6. A is (3, 2), B is (2, 3), C is (–2, –1) and D is (a, 5). Calculate the value of a if AB is parallel to CD.

7. Prove that A(–6, 3), B(2, –5) and C(–2, –1) are collinear.

8. Triangle ABC has vertices A(–5, 2), B(1, 9) and C(5, 3). Find the equation of the median AD.

SET 14

1. The line through A(2, 1) and B(p, q) is parallel to the y-axis. What can be said about p or q?
2. Calculate the area bounded by the line $y = 2x + 3$, the x-axis and the lines $x = 0$ and $x = 2$.
3. Find the equation of the straight line through the points (0, –3) and (3, 2).
4. Prove that the line joining A(–2, 1) to B(1, 3) passes through C(7, 7).
5. Find the equation of the line passing through A(2, –1) perpendicular to the line $2x - 3y = 5$.
6. Prove that the lines $y = 6$, $4y - 3x = 0$ and $x = 8$ are concurrent.
7. Triangle PQR has vertices P(4, –1), Q(7,11) and R(10, 5). Find the equation of the altitude PT.
8. Find the value of t if the lines $3x + 4y = 7$ and $2y - tx = 6$ are perpendicular.

SET 15

1. Find the equation of the line through A(0, 2) which makes an angle of 135° with the x-axis.
2. The line $ax + y + 1 = 0$ passes through the point of intersection of the line $x = 2$ and the x-axis. Find a.
3. Find the equation of the line through the point (0, 5) which is parallel to the line $3x - y = 7$.
4. Find the point of intersection of the lines $2x - y = 8$ and $3x + 2y = 5$.
5. Find the gradient of the perpendicular bisector of QR where Q is (2, –3) and R is (4, 7).
6. Triangle ABC has vertices A(1, 9), B(3, 2) and C(7, 8). Find the equation of the median AM of this triangle.
7. Find y if AB is perpendicular to CD where A is (4, 5), B(5, 3), C(4, –5) and D(3, y).
8. A, B, C and D have coordinates (2, 3), (6, 6), (4, 8) and (7, 4) respectively. Show that AB and CD are perpendicular.

SET 16

1. Calculate the angle between the line AB and the positive direction of the x-axis given that A is (–1, 2) and B is (2, 6).
2. State the gradient and intercept on the y-axis of the line with equation $\frac{x}{4} + \frac{y}{5} = 1$.
3. A is (3, 4), B(7, 2) and C(9, 1). Find the equation of the line through A parallel to BC.
4. Find the point of intersection of the lines $2x + 3y = 7$ and $5y - 3x = 18$.
5. Write down the equation of the line through the point (0, 3) perpendicular to the line $3x - 2y = 4$.
6. The points (–6, 1), (2, y) and (–2, –3) are collinear. Calculate y.
7. A, B and C are (2, 0), (3, –2) and (4, 1) respectively. Prove that ΔABC is right-angled.
8. Find the equation of the perpendicular bisector of PQ where P is (2, –1) and Q(4, –3).

SET 17

1. A point $(1, h)$ lies on the line $y = 2x + 3$. Find h.

2. Find the equation of the line passing through A(2, 4) and making an angle of 45° with the x-axis.

3. Show that the points A(–3, –9), B(2, 1) and C(5, 7) are collinear.

4. PQRS is a kite with P(6, –2) and R(3, –5). Find the gradient of QS.

5. ABCD is a parallelogram with A(1, 1), B(–1, 0) and C(–2, 3). Find the gradients of CD and AD.

6. Find the equation of the line through the point (1, –2) which is parallel to the line $3x - 2y + 1 = 0$.

7. Show that PQ is perpendicular to RS where P is (–3, –6), Q(–4, –4), R(1, 6) and S(5, 8).

8. Find the equation of the perpendicular bisector of ST where S is (1, 1) and T(2, –6).

SET 18

1. Write down the equation of the line through A(–1, 3)
   (i) and through the origin (ii) parallel to OX (iii) parallel to OY.

2. Show that PQ is parallel to RS where P is (–2, –6), Q(4, –4), R(2, 1) and S(5, 2).

3. Find the equation of the straight line through the origin perpendicular to $2x + 3y = 5$.

4. Find the point of intersection of the lines $2x + y = 0$ and $x - y = 3$.

5. Show that triangle PQR with P(1, 1), Q(4, 3) and R(1, 5) is isosceles.

6. In triangle KLM, K is (7, –1), L(–5, 5), M(–2, 1). Find the gradient of the median MP.

7. Prove that the lines $2x + 3y - 1 = 0$ and $3x - 2y - 5 = 0$ are perpendicular.

8. Find the equation of the altitude AP in triangle ABC with vertices A(–4, –3), B(–2, 3) and C(4, –1).

FUNCTIONS AND GRAPHS

SET 19

1. If $g(x) = x^4 - x^2$, evaluate $g(2) - g(-2)$.

2. On the same diagram, sketch the graphs of $y = x^2 + 2$ and $y = (x + 2)^2$.

3. State the period of (i) $\sin^2 \theta$ (ii) $\sin 2\theta$ (iii) $\sin \frac{1}{2}\theta$.

4. Calculate the exact value of $\sin^2 750°$.

5. Find the maximum value of $5 - \cos x$.

6. The functions f and g are defined on the set of real numbers by $f(x) = 2x^2 - 4$ and $g(x) = 1 - x$. Evaluate $f\big(g(2)\big)$.

7. This diagram shows part of the graph of a simple trigonometrical function.

   State the likely function.

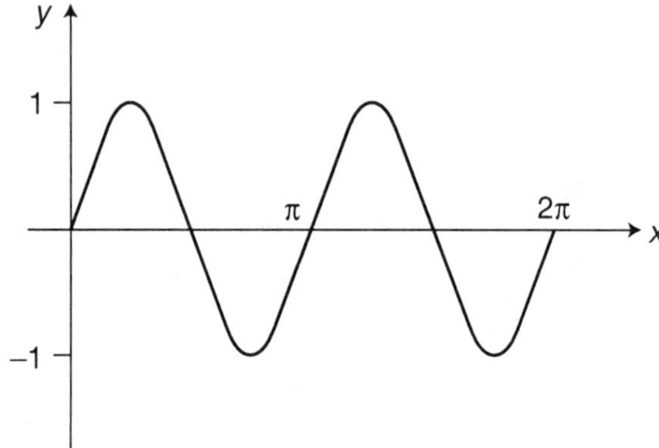

8. $f(x) = x^2 + 2x$. Find $f(x + 3)$.

SET 20

1. If $f(x) = 2x^2 - 3x + 1$, simplify $f(3x) - f(2x)$.

2. Sketch the graph of $\sin x°$ for $0 \leq x \leq 360$.

3. $f : x \to x^2 - 1$ and $g : x \to x + 3$ are mappings on the set of real numbers. Find formulae for $f\big(g(x)\big)$ and $g\big(f(x)\big)$.

4. $f(a) = 4a - 3$. Evaluate $f(1 + h) - f(1 - h)$.

5. Which elements of the domain of the function $f(x) = 7 - x^2$ map to 3?

6. A function g is defined by $g(x) = [g(x - 1)]^2$. If $g(0) = 2$, find $g(2)$.

7. Find the value of x for which $10 \sin(x - 30)°$ has its minimum value.

8. f and g are two functions defined by $f(x) = 1 - x^2$ and $g(x) = \cos x$. Find $f\big(g(x)\big)$ in its simplest form.

SET 21

1. If $f(x) = 4x^4 - 2x^2$, calculate $f(2) - f(-2)$.
2. On the same diagram sketch the graphs of $f(x) = x^2 - 2$ and $g(x) = 6 - x^2$. Write down the coordinates of the common points.
3. This diagram shows part of the graph of a simple trigonometrical function.

   State the likely function.

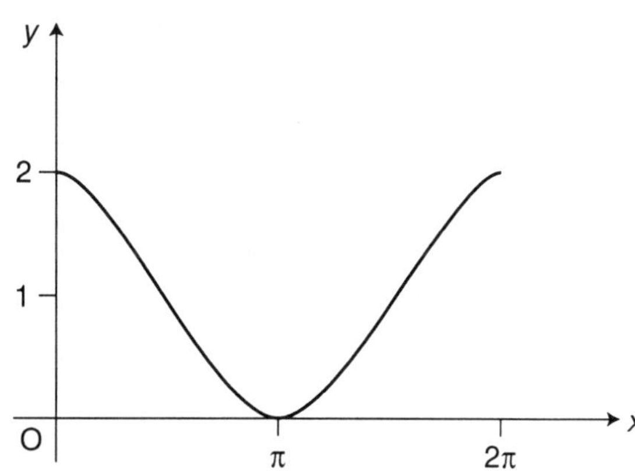

4. State the period of (i) $\cos x°$ (ii) $\cos^2 x°$.
5. If $h(x) = (2x - 1)^2$, find $h(x + 1)$.
6. State the maximum and minimum values of $4 \cos x + 1$.
7. Two functions f and g are defined as follows:
   $f(x) = x - 1$; $g(x) = (x + 1)^2 + x + 1$.
   Find in its simplest form an expression for $g(f(x))$.
8. Evaluate $\cos 450° - 2 \sin 540°$.

SET 22

1. For the function $f(x) = \sqrt{9 - x^2}$ state (i) the domain (ii) the range.
2. If $g(x) = 3x^2 + 1$, evaluate $g(x + 1) - g(x)$.
3. State the period of (i) $\sin 4x°$ (ii) $\sin \frac{1}{4}x°$.
4. Express $x^2 - 4x + 1$ in the form $(x + a)^2 + p$.
5. Write down an equation for this parabola.

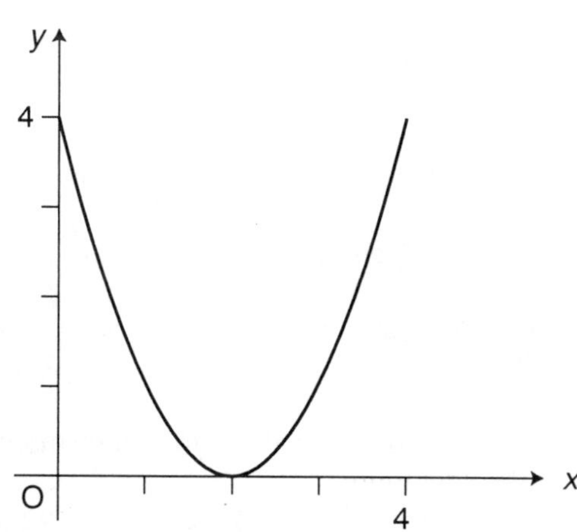

6. $f(x)$ is an **even** function if, for all x, $f(-x) = f(x)$. Which of the following are **even**:
(i) $\sin x$ (ii) $x \sin x$ (iii) $x + \sin x$?

7. $f: x \to x + 3$ and $g: x \to x^2$. Find a formula for the function $f(g(f(x)))$.

8. What is the minimum value of $k = 17 + 5 \sin (x - 35)°$?
For what value of x does it occur?

SET 23

1. Write down the exact value of
(i) $\sin \frac{\pi}{3}$ (ii) $\cos \frac{\pi}{6}$ (iii) $\tan \frac{\pi}{4}$

2. If $g(x) = (x - 1)$ and $h(x) = x^2 + x + 1$, find $h(g(x))$.

3. Find the maximum value of $g(x) = 3 - 2(x + 1)^2$ and the corresponding value of x.

4. Sketch the graph of $\sin (2x + \pi)$, $0 \le x \le \pi$.

5. $f(x) = 2 \sin x + 3$. What is the minimum value of $f(x)$?

6. If $f(x) = 2 - x^2$, simplify $f(a + 1) - f(a - 1)$.

7. State the ranges of the functions (i) x^2 (ii) x^3 (iii) $x^2 + 3$ (iv) $(x - 3)^2$.

8. The diagram shows the graph of $y = f(x)$ for $-2 \le x \le 6$.

Sketch the graph of $y = 1 - f(x)$.

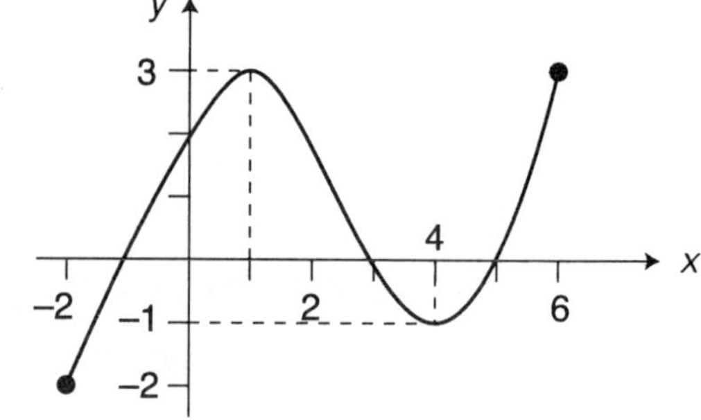

SET 24

1. Evaluate exactly $\cos 0 \,.\, \tan \frac{\pi}{6} \,.\, \sin \frac{\pi}{3}$.

2. $f(x) = 2x - 1$ and $g(x) = x^2 - 1$. Find $f(f(x))$ and $g(f(x))$.

3. Sketch the curve with equation $y = (x - 2)^3$. [Remember to indicate where it crosses the axes.]

4. $f(x) = 5 - 2 \cos x°$. What is the maximum value of $f(x)$?

5. $g(x) = x^2 + x - 1$. If $g(a) = 5$, find the values of a.

6. Write down the quadratic function represented by this graph.

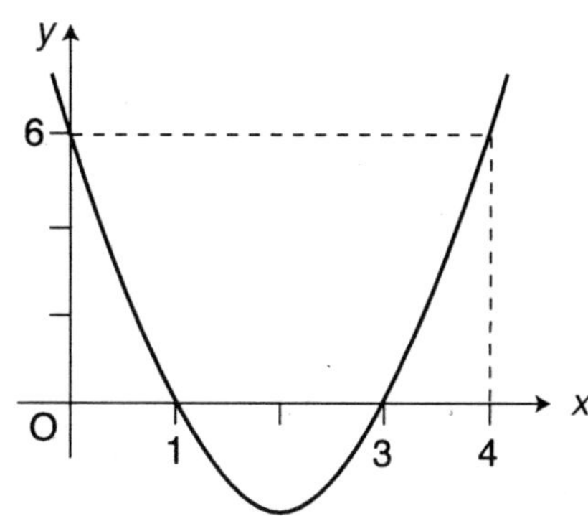

7. Express $2x^2 - 12x + 17$ in the form $k(x + a)^2 + b$.

8. Sketch the graphs of (i) $f(x) = 2^x$ (ii) $f(x) = 2^x + 1$ (iii) $f(x) = 2^{x+1}$.

DIFFERENTIATION

SET 25

1. Find the rate of change of the function $y = 3x^2$ when $x = -2$.

2. Find the derivative of $3x^2(2 - 5x)$.

3. Calculate the gradient of the tangent to the curve $y = 9 - x^2$ at the point (3, 0).

4. If $f(x) = 2x^{3/2}$, find $f'(4)$.

5. Given that $V = \pi r^2 h$, where r is constant, find the derivative of V with respect to h.

6. Calculate the angle at which the tangent to the curve $y = x^3 + x + 2$ at the point A(0, 2) is inclined to the positive direction of the x-axis.

7. $f(x) = \dfrac{x + 1}{\sqrt{x}}$, $x \neq 0$. Find $f'(x)$.

8. Differentiate $a^{1/2}(a^{1/2} - a^{-1/2})$ with respect to a.

SET 26

1. Find the rate of change of the function f given by $f(s) = 4s - s^2$ when $s = 1$.
2. If $f(x) = x\sqrt{x}$, $x > 0$, find $f'(x)$.
3. Calculate the gradient of the tangent to the curve $y = \frac{1}{3}x^3 + \frac{1}{2}x^2$ at the point where $x = -1$.
4. $y = x^3 + 1$. Show that y is never decreasing.
5. Differentiate $\frac{4x^2 + 3}{x}$; $x \neq 0$.
6. Sketch the graph of $y = f'(x)$ when $f(x) = x^2 - 2x + 3$.
7. Find the value of x for which the gradient of the tangent to the curve $y = 4x - x^2$ is 2.
8. Find the stationary point on the curve $y = x^2 - 2x$ and state its nature.

SET 27

1. Calculate the rate of change of the function f defined by $f(t) = 10t - 5t^2$ when $t = 2$.
2. Differentiate with respect to u: (i) $\sqrt{u}$ (ii) $\frac{1}{u^2}$.
3. Find the gradient of the curve $y = x(x - 2)$ at the point (3, 3).
4. Find the range of values of x for which the function f defined by $f(x) = x^2 - 4x + 3$ is increasing.
5. Evaluate $f'(1)$ where $f(z) = 4z - z^2$.
6. If $f(x) = \frac{(x + 3)(x - 1)}{\sqrt{x}}$, $x \neq 0$, find $f'(x)$.
7. If $f'(x) = (x - a)(x - b)$, state the stationary values of $f(x)$.
8. The gradient of a tangent to the curve $y = 3x^2 + 2ax + b$ at the point (1, 6) is 4. Find a and b.

SET 28

1. Calculate the rate of change of the function $f: x \to \sqrt{x}$ when $x = 4$.
2. Find $\frac{d}{dx} k(x^3 - x^2)$, ($k$ constant).
3. Calculate the gradient of the curve $y = 5x^3 - 10x$ at the point (1, –5).
4. If $f(x) = x^3 - 2x^2 + 3$, evaluate $f'(-3)$.
5. State the maximum number of stationary points on the curve $y = ax^3 - 3x^2 + 4x - 5$, where a is a constant.
6. $h(x) = \left(\sqrt{x} + \frac{1}{\sqrt{x}}\right)\left(\sqrt{x} - \frac{1}{\sqrt{x}}\right)$. Find $h'(x)$
7. Find the range of values of x for which the function g defined by $g(x) = 2x^3 - 6x$ is increasing.
8. Find the point where the gradient of the tangent to the curve $s = 5t^2$ is 10.

SET 29

1. $f(x) = 3 - 4x + 2x^2$. Calculate $f'(-2)$.

2. Differentiate with respect to p; (i) $p\sqrt{p}$ (ii) $\dfrac{p}{\sqrt{p}}$

3. Calculate the gradient of the curve $y = 3x - x^3$ at the point where $x = -2$.

4. Show that $f(x) = 6 - 2x^3$ is never increasing.

5. Find $\dfrac{dy}{dx}$ for $y = 2x(2 - 3x)$.

6. The tangent to $y = x^3 - ax^2 + 1$ at (1, 3) has gradient 5. Find the value of a.

7. $f(x) = 3(2 - \sqrt{x})^2$. Find $f'(x)$.

8. State the range of values of x for which the function $f(x) = 3x^2 - 6x + 5$ is decreasing.

SET 30

1. $A(r) = \pi r^2 + 2\pi rh$. Find $\dfrac{dA}{dr}$, where h is constant.

2. Calculate the rate of change of $g(x) = 3x^{3/2}$ when $x = 9$.

3. If $f'(x) = 3x(a - x)$, state the stationary values of $f(x)$.

4. $f(x) = (x^2 + 3)^2$. Find $f'(x)$.

5. Find the gradient of the tangent at (2, –2) on the curve $y = 3x - 2x^2$.

6. Differentiate $\dfrac{x^2 - 3x}{\sqrt{x}}$

7. The gradient of the tangent at A to the curve $y = 3x^2 - 4x + 1$ is 2. Find the coordinates of A.

8. Find the stationary point on $y = 4 - x^3$ and determine its nature.

RECURRENCE RELATIONS

SET 31

1. A sequence is defined by $u_n = \frac{1}{2}n(n + 2)$. Find u_2 and u_7.
2. Which of the sequences defined by these recurrence relations has a limit?
   (a) $u_{n+1} = 0{\cdot}6u_n + 5$ *(b)* $v_{n+1} = 6v_n + 0{\cdot}5$
   Explain.
3. A sequence is defined by $u_n = 3 \,.\, 4^n$. Evaluate $\frac{u_{n+1}}{u_n}$.
4. Find the value of n if $u_n = 16$ in the sequence defined by $u_n = \frac{n(n-2)}{3}$.
5. Calculate exactly the limit of the sequence defined by the recurrence relation $u_{n+1} = 0{\cdot}4\,u_n + 1{\cdot}2$.

6. A sequence is defined by the recurrence relation $u_{n+1} = 3u_n + 4$ and $u_0 = -1$. Which is the first term of this sequence to exceed 100?
7. The two sequences defined by the recurrence relations $u_{n+1} = 0{\cdot}3u_n + 14$ and $v_{n+1} = 0{\cdot}4\,v_n + k$ have the same limit. Find the value of k.

8. Medication is given to a patient at the rate of 50 grams every six hours. If the amount in his body decreases by 20% each hour, find how much remains immediately before the second dose is given.

SET 32

1. The nth term of a sequence is defined by $u_n = \frac{n}{2}[1 - (-1)^n]$. Write down u_{15} and u_{16}.
2. A sequence is defined by $u_n = \frac{1}{2}n(n + 2)$. Find n when $u_n = 24$.
3. Calculate exactly the limit of the sequence defined by the recurrence relation $u_{n+1} = 0{\cdot}7u_n + 6$.
4. The sequence defined by $u_{n+1} = au_n + 32$ has a limit of 40. Find the value of the constant a.
5. For the sequence defined by $u_n = 3n + 5$, evaluate $u_{n+1} - u_n$.

6. Find the first term to exceed 10 in the sequence defined by $u_{n+1} = 0{\cdot}8u_n + 3$ where $u_1 = 5$.
7. The two sequences defined by the recurrence relations $u_{n+1} = 0{\cdot}1u_n + p$ and $v_{n+1} = 0{\cdot}7v_n + q$ have the same limit. Express p in terms of q.
8. The manufacturer claims that spraying weekly with a pesticide will kill 70% of the pests. However, 600 new pests will invade the bushes between sprayings. Express as a recurrence relation.

ALL TOPICS IN UNIT 1

SET 33

1. The line $x + 2y = 6$ cuts the x-axis at A and the y-axis at B. Find the mid-point of AB.

2. Find the value of the derivative of the function f defined by $f(x) = 2x^2 - 3$ at the point where $x = 5$.

3. Find the equation of the line through $(2, -1)$ perpendicular to the line $x + 3y = 4$.

4. If $A = 2\pi r(r + h)$, where h is constant, find the derivative of A with respect to r.

5. $f(x) = 3x^2$ and $g(x) = 2x + 1$. Find formulae for (i) $f(g(x))$ (ii) $g(f(x))$.

6. Prove that triangle ABC is right angled and isosceles where A is (–3, 8), B is (–1, 2) and C is (5, 4).

7. A sequence is defined by $u_n = 4n + 7$. Evaluate $u_{n+1} - u_n$.

8. $f(x) = \left(\sqrt{x} + \dfrac{2}{\sqrt{x}}\right)^2$, $x \neq 0$. Find $f'(x)$.

SET 34

1. Differentiate $\dfrac{1}{u^{1/2}}$ with respect to u.

2. Find z if PQ is perpendicular to RS where P is (4, 2), Q(6, 4), R(0, –3) and S(2, z).

3. Calculate exactly the limit of the sequence defined by the recurrence relation $u_{n+1} = 0{\cdot}2u_n + 5$.

4. Write down an expression for the rate of change of $f(x)$ when $x = a$.

5. For what values of x is $f(x) = 3x^4 - 4x^3$ increasing?

6. Find y if A is (–7, 2), B(8, y) and AB = 17.

7. Express $6x - x^2 - 7$ in the form $a - (x - b)^2$.

8. Triangle OAB has vertices O(0, 0), A(7, 5) and B(1, 11). Find the equation of
(i) the altitude OF of ΔOAB (ii) the median OQ of ΔOAB.

SET 35

1. Calculate the gradient of the tangent to the curve $y = 2x^2 - 1$ at the point where $x = 0$.
2. The line $y - mx = 4$ is perpendicular to the line $2y + x = 3$. Find m.
3. $f(x) = \dfrac{2^x}{2^{2x} + 1}$. Evaluate $f(-1)$.
4. (i) $f(x) = 3 - 4x - x^2$; find $f'(-1)$.
   (ii) $f(z) = 3 - 4z - z^2$; find $f'(-1)$.
   (iii) $g(s) = 3 - 4s - s^2$; find $g'(-1)$.

5. Find the first five terms of the sequence defined by $u_{n+1} = 10u_n - 9$ when $u_1 = 2$.
6. Find the stationary point on the curve $y = x^3$ and determine its nature.
7. Find k if the lines $2x - y = 5$, $3x - 2y = 9$ and $4x + ky = 1$ are concurrent.
8. A(3, –1), B(–1, 4) and C(5, 2) are the three vertices of a triangle. Calculate:
   (i) the gradient of median AD (ii) the gradient of altitude BE.

SET 36

1. Write down the gradient of all lines perpendicular to the join of A(2, 5) and B(2,14).
2. Find the point of intersection of the lines $2x + 4y = 10$ and $3x - 2y = 25$.
3. Find the equation of the straight line through (0, 2):
   (i) parallel to $3x - y = 7$ (ii) perpendicular to $5x - 4y = 10$.
4. Write down an expression for the gradient of the tangent to the curve $y = f(x)$ at the point $(a, f(a))$.
5. Evaluate $f(4)$ where $f(x) = x^{3/2} \cdot \left(x - \frac{3}{2}\right)$
6. Differentiate $u^{1/2} - 3u^{3/2}$ with respect to u.
7. State the range of values of x for which the function f defined by $f(x) = x^3 - 3x$ is decreasing.
8. A sequence is defined by $u_n = n^2 - n + 1$.
   Find (i) u_{10} (ii) n when $u_n = 73$.

UNIT 2

REVISION OF STANDARD GRADE ESSENTIALS AND UNIT 1

SET 37

1. Solve the equation $(x-1)(x-2) = 6$, where x is a real number.

2. $f(x) = 1 - x^2$ and $g(x) = 2x - 1$. Find a formula for $f(g(x))$ and factorise it.

3. Sketch the graph of $\cos(\pi + x)$ for $0 \leq x \leq 2\pi$.

4. The gradient of the curve $y = 3x - x^2$ at point A is -1. Find the coordinates of A.

5. Find the equation of the line through the origin perpendicular to the line $3y - 4x + 2 = 0$.

6. Simplify $\dfrac{a^2 - 7a + 12}{a^2 - 9} \times \dfrac{a^2 + 2a - 3}{a^2 - 8a + 16}$

7. $f: x \rightarrow x^2$ and $g: x \rightarrow x - 1$. Find the derivative with respect to x of
(i) $f(g(x))$ (ii) $g(f(x))$

8. Prove that the lines $x - y + 4 = 0$, $x - 2y + 7 = 0$ and $3x - y + 6 = 0$ are concurrent, stating the point of intersection.

SET 38

1. Solve the equation $x^2 + x = (x+1)^2$

2. Find the values of x for which the curve $y = x^3 + x^2 - x + 2$ has stationary points.

3. Calculate exactly the limit of the sequence defined by the recurrence relation $u_{n+1} = 0{\cdot}6u_n + 5$.

4. Simplify $\dfrac{1}{(x+2)^2} - \dfrac{1}{x^2}$

5. Find the exact length of BC in triangle ABC.

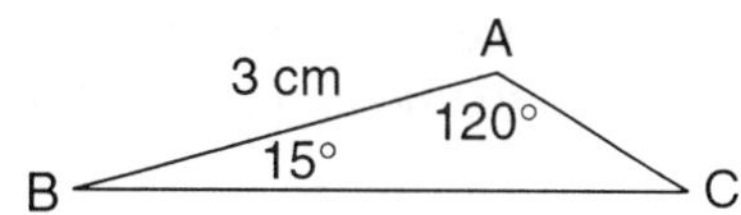

6. $f(x) = x^2$. Sketch the graphs of (i) $f(x)$, (ii) $f(x) + 1$, (iii) $f(x+1)$, (iv) $-f(x+1)$, (v) $f'(x)$, showing turning points and intersections with coordinate axes.

7. $f(x) = x^2 + x$. Find $f(f'(x))$ where f' is the derivative of f.

8. Triangle PQR has vertices P(1, 7), Q(–1, 2) and R(2, 4). Find the equation of the altitude PS of this triangle.

SET 39

1. If $\frac{1}{u} + \frac{1}{v} = \frac{1}{f}$ $(u, v, f \neq 0)$, find f in terms of u and v.

2. Differentiate with respect to z, $f(z) = \dfrac{z^2 + 1}{\sqrt{z}}$

3. Give the maximum and minimum values of $6 - 2\cos x°$.

4. A sequence is defined by $u_n = n^2 + 3n - 2$. Find (i) u_7, (ii) n when $u_n = 52$.

5. Simplify (i) $(2\sqrt{3} - 5)(2\sqrt{3} + 5)$ (ii) $\dfrac{\sqrt{50} - \sqrt{75}}{\sqrt{27} - \sqrt{18}}$

6. Find the equation of the median LN in triangle KLM with K(2, 4), L(–1, 1) and M(3, –1).

7. Find the coordinates of the points at which the tangent to $y = x^3 + 3x^2 - 2x + 5$ is perpendicular to the line $2x - 4y = 5$.

8. (i) Sketch the graph of the function defined by $f(x) = \begin{cases} x + 2 : x \leq 0 \\ 2 - x : x > 0 \end{cases}$

   (ii) Explain why f does not have an inverse.

SET 40

1. Simplify $\left(x - \frac{1}{x}\right)\left(x^2 + 1 + \frac{1}{x^2}\right)$, $x \neq 0$.

2. If $f(x) = (x + 4)^2$, find $f'(x)$.

3. Find the cosine of the smallest angle in a triangle with sides measuring 3 cm, 5 cm and 6 cm.

4. If $g(x) = 3x^2 - 2x$, find in factorised form $g(3 - 2x)$.

5. Solve (i) $5^x = \frac{1}{625}$ (ii) $4x^{2/3} - 1 = 35$

6. This is the graph of a cubic function $y = f(x)$. Sketch the graph of $y = f'(x)$.

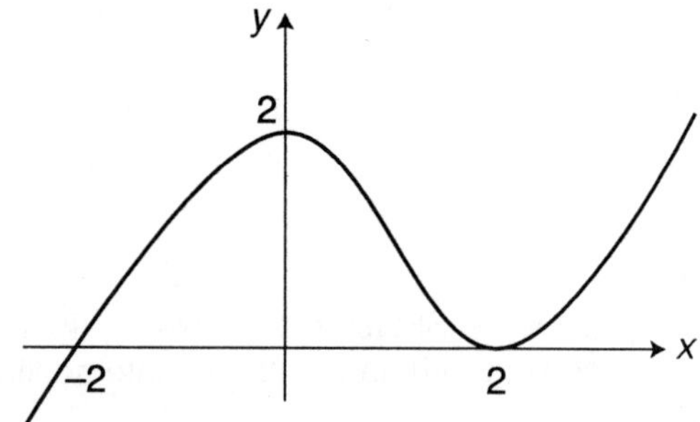

7. Find the equation of the tangent to the curve $y = x^3 - 3x^2 + 1$ at the point where $x = -1$.

8. If PQ is perpendicular to RS and P is (5, –7), Q(–1, –3), R(a, 7) and S(6, 10), find the value of a.

SET 41

1. Solve $3x - 2y = 8$
   $2x - 3y = 7$

2. If $f(x) = 2 - 5x^2 + x^3$, evaluate $f'(-1)$.

3. This is the graph of $f(x) = a + b \sin cx$ for $0 \leq x \leq 180$.

   Find the values of a, b and c.

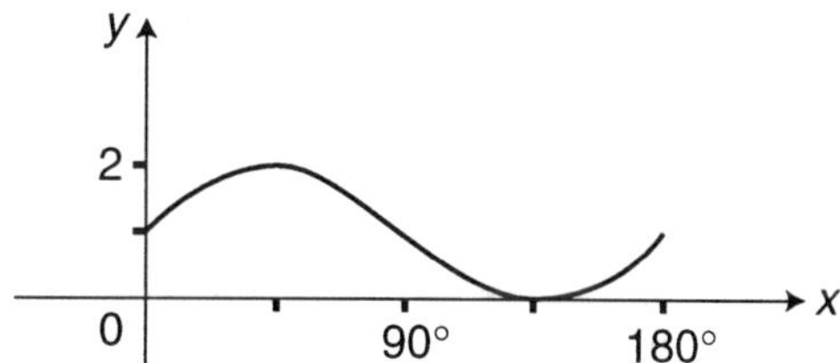

4. If $f(x) = 1 - x^2$ and $g(x) = (1 - x)^2$, find in its simplest form an expression for $g(f(x))$.

5. For the points P(1, –1), Q(–2, 2) and R(1, k), find the values of k so that PQ = QR.

6. A sequence is defined by $u_{n+1} = 2u_n + 5$, $u_0 = 3$. Find the first term to exceed 250.

7. A ball is thrown vertically upwards reaching a height of h metres after t seconds where $h = 1 + 20t - 4t^2$. Find the speed of the ball after 2 seconds and the time taken to reach the maximum height.

8. The equations of the three sides of triangle KLM are KL: $3x - y + 4 = 0$, LM: $3x + y + 5 = 0$, MK: $x - 3y - 1 = 0$. Show that this triangle is right angled and name the right angle.

SET 42

1. Simplify $4x(3x - 7) - 3x(2 - x) - x$, giving the answer in factorised form.

2. Differentiate with respect to u, $f(u) = 3u^{3/2} + 2u^{1/2}$.

3. If $3 - 2\cos^2 x° = 1$, find x. $(0 \leq x \leq 360)$

4. Find the value of a if the sequence defined by the recurrence relation $u_{n+1} = 0{\cdot}7u_n + a$ has a limit of 5.

5. Points P(–2, 5), Q(1, c) and R(7, 2) are collinear. Find the value of c.

6. Write down the quadratic function represented by this graph.

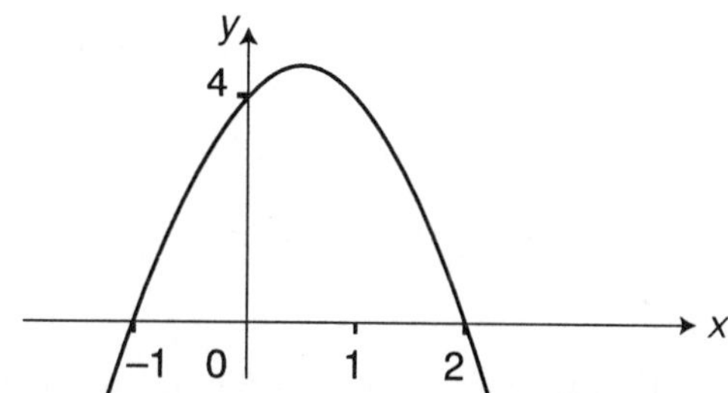

7. Find the coordinates of the points on the curve $y = x^3 + 4x^2$ at which the gradient is 3.

8. Find the equation of the altitude from T in triangle STV with vertices S(–2, –2), T(2, 1) and V(4, –3).

SET 43

1. Solve $\frac{(x+3)}{2} = 1 - \frac{(x-2)}{3}$

2. Find the gradient of the curve $f(x) = 2x^3 - 5x^2 + 1$ at the point $(2, -3)$.

3. Write down the maximum and minimum values of $3 \sin x° - 2$.

4. A sequence is defined by $u_n = 3 + 5n - n^2$. Find (i) u_6, (ii) n, when $u_n = -21$.

5. If $g(x) = \frac{1}{x-1}$ and $h(x) = x^2$, write down an expression for $g(h(x))$.

6. From the graph of $f(x)$, sketch the graphs of
   (i) $f(x) - 2c$
   (ii) $-f(x)$.

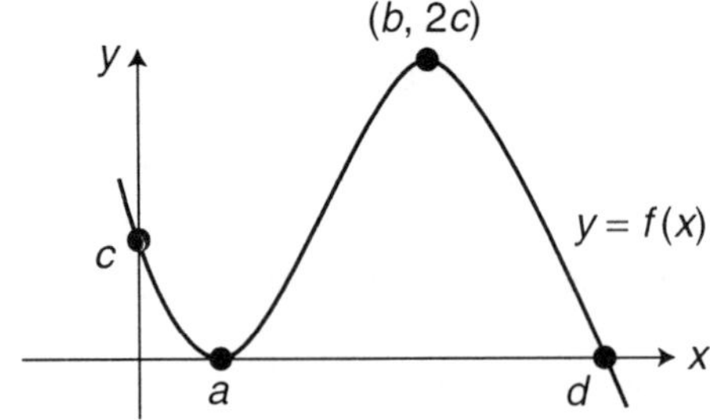

7. Find the stationary points on the curve $y = x^3 - 3x$ and determine their nature.

8. ABC is a triangle with vertices A(0, 5), B(–2, 3) and C(4, –1). Find the equation of the line through the mid-point of AB parallel to BC.

SET 44

1. Factorise fully (i) $3x^4 - 12x^2$ (ii) $5 \cos^2 x° - 4 \cos x° - 1$.

2. Find $\frac{dy}{dx}$ when $y = \frac{(1-2x)^2}{x}$

3. Evaluate sin 45° cos 120° – cos 45° tan 120° leaving the answer in surd form with a rational denominator.

4. Calculate the area of the triangle bounded by the lines $y = x$, $y = 9 - 2x$ and the y-axis.

5. Express $x^2 + 5x - 1$ in the form $(x + a)^2 + p$.

6. Find the equation of the tangent to the curve $y = x^3 - 2x + 2$ at the point where $x = 2$.

7. In triangle OEF, with E(–3, 0) and F(0, 6), the medians intersect at M. Find the coordinates of M.

8. Eagles eat 10% of the rabbits in a warren every day. Fifteen rabbits are born every night. What will happen to the population of rabbits in the long term?

THE REMAINDER THEOREM AND QUADRATIC THEORY

SET 45

1. State the degree of the polynomial $x - 2x^2 + x^3 - x^4$ and give the coefficient of x^2.
2. Calculate the remainder when $x^3 - 11x + 10$ is divided by $x + 3$.
3. Find the remainder and quotient when $4x^3 - 2x^2 + 4x - 5$ is divided by $x - 2$.
4. For what value of k is $(x + 2)$ a factor $9x^3 + kx^2 - x - 2$?
5. Find p if the remainder on dividing $x^4 + px^3 - 5x + 11$ by $x - 3$ is -4.
6. Factorise fully $x^3 - 3x^2 - 4x + 12$.
7. Factorise fully $x^3 - x^2 - 2x$.
8. If $x - 1$ is a factor of $x^3 + a^2x^2 + 2a$, find the value of a.

SET 46

1. Find the remainder on dividing $1 - x^3$ by $x - \frac{1}{2}$.
2. Factorise fully $x^3 + 4x^2 - 7x - 10$.
3. Factorise fully $x^3 - 7x + 6$.
4. Show that the equation $x^3 - 6x + 1 = 0$ has a root between 0 and 1.
5. Solve $x^3 + 2x^2 - x - 2 = 0$.
6. Find the value of q if the remainder is -3 when $x^3 + 2x^2 - 11x + q$ is divided by $x + 5$.
7. Solve $2x^3 - 19x^2 + 59x - 60 = 0$.
8. Solve $3x^3 + 4x^2 - 13x + 6 = 0$.

SET 47

1. If $x^2 + ax + 9 = (x + p)^2$, find a and p (both >0).
2. If $(x + k)^2 = x^2 - 3x + m$, find the values of k and m.
3. Write down the equation of the axis of symmetry of the parabola with equation $y = 3(x - 2)(x + 4)$.
4. Find the solution set of the inequality $x^2 - 3x + 2 > 0$, where x is a real number.
5. Solve $6x^2 - 11x - 10 \geq 0$.
6. Solve $(x + 1)^2 < 9$.
7. Express $x^2 + 6x + 8$ in the form $(x + a)^2 + b$.
8. Express $7 - 2x - x^2$ in the form $p - (x + q)^2$.

SET 48

1. State the nature of the roots of the equation $2x^2 - 5x - 7 = 0$.
2. Determine the nature of the roots of the equation $(x - 2)(x - 1) = 30$.
3. Find p if $x^2 + p = 8x$ has equal roots.
4. State the condition for the equation $x^2 + 2bx + a^2 = 0$ to have equal roots.
5. The equation $x^2 = k(2x - 5)$ has equal roots. Find k.
6. For what range of values of k does the equation $3x^2 + 4x + k = 0$ have real roots?
7. Show that the roots of $5x^2 - kx - 1 = 0$ are always real.
8. Show that for all values of k the equation $k^2x^2 - 2kx + 1 = 0$ has equal roots.

SET 49

1. Find the remainder on dividing $13 + 7a - 5a^3$ by $a + 4$.
2. The remainder on dividing $x^3 + kx + 7$ by $x + 2$ is 3. Find k.
3. $4x^3 + 4x^2 - 11x + p$ is divisible by $x + 2$. Find p.
4. Determine the nature of the roots of the equation $(2 - x)(3 - x) + 1 = 0$.
5. Show that $3x^3 - 2x + 5 = 0$ has a root between -2 and -1.
6. State the values of k for which the equation $k^2x^2 + 2x + 1 = 0$ has equal roots.
7. For what values of k does the equation $kx^2 + 2x + (2k - 1) = 0$ have equal roots.
8. Show that if the roots of $7x^2 - px + 1 = 0$ are real then the roots of $8x^2 + 2px + 3 = 0$ are also real.

SET 50

1. Find the value of k if $(x + 3)$ is a factor of $2x^3 - 5x^2 + kx + 18$.
2. Find a real root of the equation $2x^3 - 3x^2 + 2x - 8 = 0$.
3. For what value of the constant a does the equation $ax^2 + 8x + 16 = 0$ have equal roots?
4. Show that $(x - 3)$ is a factor of $g(x) = 4x^3 + 6x^2 - 46x - 24$, and express $g(x)$ in its fully factorised form.
5. How many real roots has the equation $(x - 1)(x^2 - x + 7) = 0$?
6. Find the quotient and remainder when $x^3 - 4x + 10$ is divided by $(x - 2)$.
7. Find the remainder on dividing $x^{101} + 101$ by $x + 1$.
8. Rearrange the equation $k = \dfrac{a^2 + 3a + 4}{a + 2}$, $a \neq -2$, as a quadratic in a.

INTEGRATION

SET 51

1. If $f(x) = x^2$, find $f'(x)$ and $\int f(x)\,dx$.

2. Find $f(x)$ given that $f'(x) = 4x^3 - 2x + 1$.

3. If $\frac{dy}{dx} = 1$, find y, given that $y = 0$ when $x = 0$.

4 $\frac{ds}{dt} = 3t^2 - 1$. Find an expression for s given that $s = 3$ when $t = 2$.

5. $\frac{dy}{dx} = 2x + 1$ and $y = 3$ when $x = 1$. Find an expression for y.

6. Find the anti-derivative of $4 - 2x$.

7. Find the equation of the graph of $y = f(x)$ where $\frac{dy}{dx} = 2$ and $f(2) = 0$.

8. If $g(x) = (3x - 1)^2$, find (i) $g'(x)$ (ii) $\int g(x)\,dx$.

SET 52

1. If $f(x) = (3 - x)(3 + x)$, find (i) $f'(x)$ and (ii) $\int f(x)\,dx$.

2. Find $\int (x + 2)^2\,dx$.

3. $f'(x) = x(2x^2 - 1)$. Find $f(x)$.

4. Integrate $(2x - 1)^2$ with respect to x.

5. Find $f(x)$ given that $f'(x) = \frac{1}{x^2}$, $x \neq 0$.

6. Evaluate $\int_{-5}^{5} x\,dx$.

7. Illustrate by a sketch $\int_{-1}^{2} x^2\,dx$.

8. Given that $\int_a^b f(x)\,dx = k$, write down the value of $\int_b^a f(x)\,dx$.

SET 53

1. Calculate the shaded area shown.

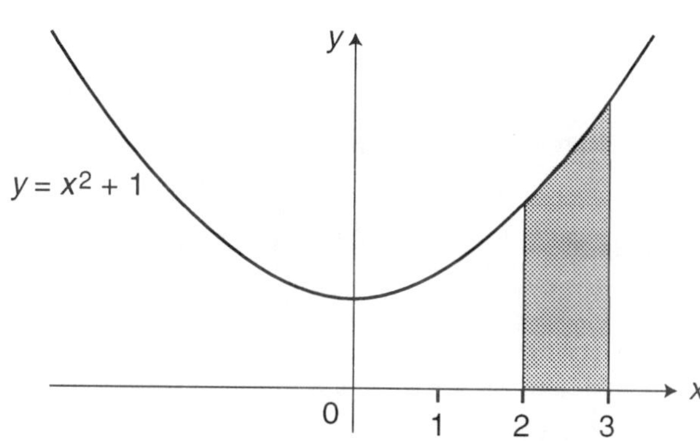

2. Evaluate $\int (6x^2 - 2x)\,dx$

3. Evaluate $\int_{-1}^{2} x(x - 3)\,dx$

4. Evaluate $\int_{-1}^{2} (x - 3)^2\,dx$

5. Find $\int f(x)\,dx$, where $f(x) = (2x + 1)^2$.

6. Write down an expression for the shaded area.

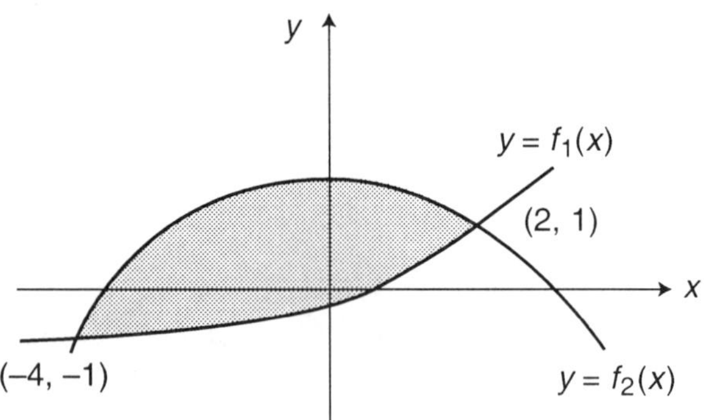

7. Given that $\frac{ds}{dt} = 3t^2 - 2t$, find an expression for s given that $s = 4$ when $t = 1$.

8. The curve for which $\frac{dy}{dx} = 6x^2 - 2x$ passes through the point (–1, 2). Express y in terms of x.

SET 54

1. Evaluate $\int_{-1}^{1} x^4\,dx$

2. Evaluate $\int_{-1}^{1} x(x^2 - 2)\,dx$

3. Given that $a > 1$ and $\int_{1}^{a} (2x + 3)\,dx = 24$, find a.

4. Show in a sketch the area given by $\int_{-2}^{2} x^3\,dx$. Hence state the value of $\int_{-2}^{2} x^3\,dx$.

5. Sketch the graph of the function f given by $f(x) = x^3$. Given that $\int_{0}^{a} x^3\,dx = \frac{1}{4}$, use your graph to evaluate $\int_{-a}^{0} x^3\,dx$.

6. Illustrate the integral $\int_{-1}^{2} (4 - x^2)\,dx$ by means of a sketch.

7. Evaluate $\int_{1}^{8} \sqrt[3]{x^2}\,dx$

8. Evaluate $\int_{0}^{4} (x^{3/2} + x^{1/2})dx$.

SET 55

1. If $\int_c^d g(x)\,dx = k$ and $\int_d^e g(x)\,dx = z$, find $\int_c^e g(x)\,dx$.

2. Integrate with respect to x (i) $2x + 1$ (ii) $x(2x + 1)$.

3. Evaluate $\int_{-1}^{1} (2x + 1)\,dx$.

4. Given that $\int_1^a (3x - 1)\,dx = 2$ and $a > 1$, find the value of the constant a.

5. Evaluate $\int_1^9 \sqrt{x}\,dx$.

6. The graph of $y = f(x)$ crosses the y-axis at (0, 4). Given that $f'(x) = 2x - 3$, find $f(x)$.

7. Evaluate $\int_{-1}^{2} (x + 3)^2\,dx$.

8. Write down an integral which is a measure of the shaded area.

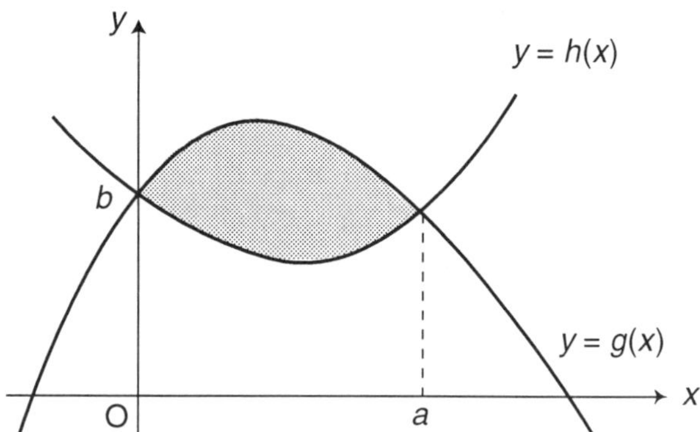

SET 56

1. Simplify $x^{3/2}(x^{1/2} + x^{-1/2})$ and hence find $\int x^{3/2}(x^{1/2} + x^{-1/2})\,dx$.

2. Find $\int (2x^3 - 3x^2 + x - 1)\,dx$.

3. Find $f(x)$, given that $f'(x) = 1 - \frac{1}{x^2}$, $x \neq 0$.

4. The parabola with equation $y = (2 - x)(3 - x)$ cuts the x-axis at A and B. Write down the coordinates of A and B.

5. Write down an expression for the area enclosed by the parabola with equation $y = (2 - x)(3 - x)$ and the x-axis.

6. Express y in terms of x, given that $\frac{dy}{dx} = 2x + 3$.

7. A curve for which $\frac{dy}{dx} = 2x + 3$ passes through the point (1, 5). Express y in terms of x.

8. Sketch the graph of $x^2 + y^2 = 4$. Shade the region which represents $\int_0^2 \sqrt{4 - x^2}\,dx$.

TRIGONOMETRY

SET 57

1. Find the solution set of the equation $3 \cos x° + 2 = 0$, $90 \le x \le 270$.
2. Find the solution set of the equation $\cos (x - 16)° = -\frac{\sqrt{3}}{2}$, $0 < x \le 360$.
3. Solve $2 \sin x \cos x = \sin x$, $0 \le x \le \pi$.
4. Simplify (i) $\sin^3 x + \sin x \cos^2 x$ (ii) $\sin^2 A + \sin^2 (90° - A)$.
5. Simplify (i) $2 \cos^2 A - \cos 2A - 1$ (ii) $\cos^2 3A + \sin^2 3A$.
6. ABCD,EFGH is a cube. Calculate the size of the angle between the base ABCD and the line DG.

7. ABCD,PQRS is a cuboid with AB = 12 cm, AD = 4 cm and AP = 3 cm. Calculate the length of a space diagonal and the angle it makes with base ABCD.

8. ABCDE is a right square pyramid with AB= 6 cm and EC = 5 cm.

   Calculate the size of the angle between planes EBC and ABCD [i.e. $E\hat{M}X$].

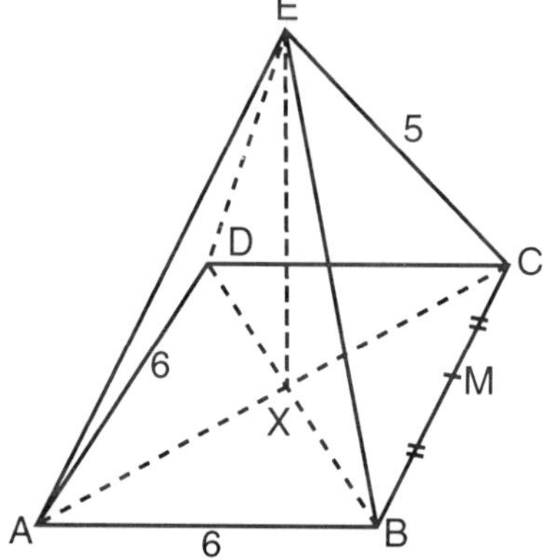

SET 58

1. Find the value of $(3 \sin x - \cos x)^2 + (\sin x + 3 \cos x)^2$.
2. Given that $\tan a° = \frac{1}{5}$, find the **exact** value of $\cos 2a°$.
3. Simplify (mentally) $\sin (x + 2y) \cos (y - x) + \cos (x + 2y) \sin (y - x)$.
4. PQRS,TUVW is a cube of side 4 cm. M and N are the mid-points of UV and VW respectively. Calculate:
   (i) the length of PM (ii) the size of $P\hat{M}N$.
5. Simplify
   (i) $\cos 45° \cos 20° - \sin 45° \sin 20°$
   (ii) $\sin 90° \cos 45° + \cos 90° \sin 45°$
   (iii) $\sin 40° \sin 30° - \cos 40° \cos 30°$.
6. By expressing 75° as (45° + 30°), calculate the exact value of cos 75°.
7. Find the solution set of the equation $\cos 2x° = \frac{\sqrt{3}}{2}$, $0 \le x < 360$.
8. If α and β are acute angles and $\tan \alpha = \frac{1}{2}$, $\tan \beta = \frac{1}{3}$, calculate the **exact** size of $(\alpha + \beta)$.

SET 59

1. Simplify:
   (i) $\cos 7x \cos 5x + \sin 7x \sin 5x$;
   (ii) $\cos^2 5x + \sin^2 5x$.

2. Factorise fully $\cos 2x + \cos x$.

3. Simplify $\sin(60° + x) - \sin(60° - x)$.

4. Simplify as far as possible $\cos^2 A + \sin^2 A - 2\sin^2 \frac{A}{2}$.

5. Simplify (mentally) the expression $\sin(A + B) . \cos(B + C) - \cos(A + B) . \sin(B + C)$.

6. ABCD,EFGH is a cuboid with AB = 5 cm, AD = 3 cm and AE = 2 cm. M is the mid-point of BC. Calculate the angle between planes EHM and EFGH ($B\hat{E}F$).

7. E,ABCD is a right square pyramid with AB = 6 cm and vertical height 9 cm. Calculate the size of $D\hat{E}B$.

8. P,QRST is a right rectangular pyramid with height 10 cm. The diagonals of the base measure 10 cm. Calculate the angle between PT and the base QRST.

SET 60

1. Evaluate $(2\sin y - 5\cos y)^2 + (5\sin y + 2\cos y)^2$.

2. Find the exact solutions of the equation $\sin(x - 50)° = \frac{1}{2}$, $0 \le x \le 360$.

3. Simplify $\sqrt{2}\left[\sin\left(\theta + \frac{\pi}{4}\right) - \cos\left(\theta + \frac{\pi}{4}\right)\right]$.

4. Find the exact solutions of the equation $4\sin^2 x° = 1$, $0 \le x \le 360$.

5. Find the exact solutions of $\sin x° \cos 30° + \cos x° \sin 30° = \frac{\sqrt{3}}{2}$ for $0 \le x \le 360$.

6. Solve the equations
   (i) $3\cos^2 x° - \cos x° = 0$
   (ii) $\sin^2 x° - 3\sin x° + 2 = 0$, $0 \le x \le 360$.

7. Solve $\sin 2x° = 4\sin x°$, $0 < x \le 360$.

8. Rectangle ABCD is vertical and isosceles triangle ABE is horizontal. AB = 10 cm, AD = 4 cm, AE = 13 cm. Calculate the angle between planes DCE and ABCD ($E\hat{P}Q$).

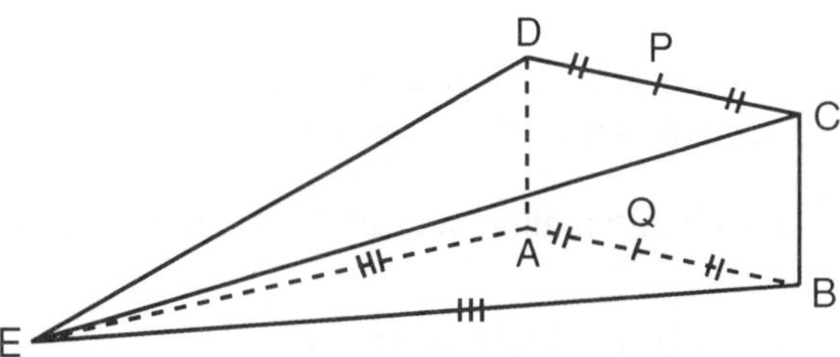

SET 61

1. Simplify $\sin(A - B) + \sin(A + B)$.

2. Simplify (i) $2\cos^2 A - 1 - \cos 2A$ (ii) $2\sin 2A\cos 2A + \sin(-4A)$.

3. Simplify $\cos^2 2A + \sin^2 2A - 2\sin^2 A$.

4. Find the exact value of x if $\sin 55^\circ \cos x^\circ - \cos 55^\circ \sin x^\circ = \frac{1}{2}$, $0 \le x \le 180$.

5. Factorise $\cos^4 A - \sin^4 A$ and simplify it.

6. PQRS,WXYZ is a cube. Calculate the size of the angle between planes PQRS and PRYW.

7. ABCD,EFGH is a cuboid with AB = 5 cm, BC = 10 cm and AE = 7 cm. Calculate angle FAB, between plane AFGD and the base ABCD.

8. ABCD,EFGH is a cube of side 2 cm. Calculate the angle between AG and the base ABCD.

SET 62

1. If $\sin x^\circ = \frac{4}{5}$, $0 < x < 90$, calculate the exact value of $\sin 2x^\circ$.

2. E,ABCD is a right square pyramid with AB = 8 and EC = 10.

   Calculate the angle between EC and the plane ABCD ($E\hat{C}A$).

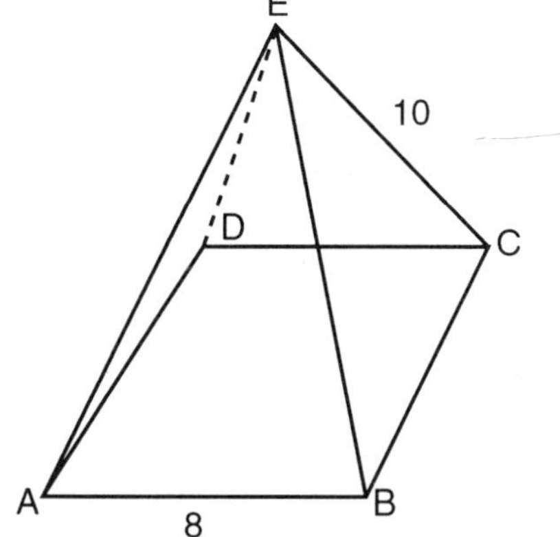

3. Evaluate exactly $(\sin 63^\circ \cos 27^\circ + \cos 63^\circ \sin 27^\circ)^2$.

4. If the exact value of $\cos x$ is $\frac{1}{\sqrt{5}}$, find the exact value of $\cos 2x$.

5. V,ABCD is a right rectangular pyramid with AB = 8 cm, BC = 6 cm and height 5 cm. Calculate angle VAC, between AV and the base ABCD.

6. Solve the equation $\tan^2 2x^\circ = 1$, given $0 \le x < 180$.

7. A and B are acute angles such that $\tan A = \frac{5}{12}$ and $\tan B = \frac{3}{4}$. Find the exact value of $\sin(A - B)$.

8. Simplify $\cos A \sin B - \sin A \cos B - \sin(B - A)$.

THE CIRCLE

SET 63

1. $(h, -1)$ lies on circle $x^2 + y^2 = 10$. Find h.

2. Find the solution set of the system of equations $\left.\begin{array}{r} y - x = 0 \\ x^2 + y^2 = 2 \end{array}\right\}$

3. State whether the point (4, 6) lies **inside**, **outside** or **on** the circle $(x - 1)^2 + (y + 2)^2 = 25$.

4. P(2, –3) and Q(6, 7) are two diametrically opposite points on a circle. State its centre and radius.

5. Find the radius of a circle which passes through (0, 1) and (0, 3) and touches the x-axis.

6. Find the equation of the circle which passes through the points (0, 1), (0, 5), (1, 0), (5, 0).

7. Write down the equation of the circle which is the image of the circle $(x - 1)^2 + (y - 2)^2 = 9$ under reflection in the x-axis.

8. P(–4, 3) is a point on the circle $x^2 + y^2 = 25$. Calculate the gradient of the tangent at P.

SET 64

1. AB is a diameter of the circle centre (1, –3). If A is (–2, –5), find the coordinates of B.

2. P(–2, –3) and Q(6, –3) are opposite ends of a diameter of a circle. Calculate the centre and radius.

3. Find the equation of the circle $x^2 + y^2 = 9$ after it is moved four units up the y-axis.

4. Find the equation of the circle passing through the origin with radius $\sqrt{8}$ and centre on the line $y = x$. (Two possible answers.)

5. State the coordinates of the centre and the length of the radius of the circle $x^2 + y^2 - 4x + 2y - 11 = 0$.

6. Find in its general form the equation of the circle with centre (3, 1) and radius 4.

7. Find the equation of the tangent to the circle $x^2 + y^2 = 25$ at the point (–4, 3).

8. Two circles are drawn so that each passes through the centre of the other. The line joining their centres is parallel to the x-axis. The equation of the left-hand circle is $x^2 + y^2 - 6x - 4y - 3 = 0$. Find the equation of the other circle.

SET 65

1. Explain why the equation $x^2 + y^2 + 2x + 4y + 6 = 0$ does not represent a circle.

2. The circle $x^2 + y^2 - 3x + 7y - 10 = 0$ cuts the x-axis at A and B. Find the length of the chord AB.

3. State the centre and radius of the circle with equation $x^2 + y^2 - 4x + 6y - 3 = 0$.

4. Give the coordinates of the centre of the circle $2x^2 + 2y^2 - 4x + 7y - 1 = 0$.

5. Find the equation of the image of the circle $x^2 + y^2 - 2x - 4y - 4 = 0$ under reflection in the y-axis.

6. Write down the coordinates of the centre of the circle $3x^2 + 3y^2 - 6x - 12y - 19 = 0$.

7. The circle $x^2 + y^2 - 4x + 2y - 11 = 0$ maps on to circle $x^2 + y^2 = 16$ under a translation. Describe it.

8. Find the equation of the circle which touches the y-axis, and touches the line $y = -2$ at the point $(5, -2)$.

SET 66

1. Find the gradient of the tangent to the circle $(x - 1)^2 + (y - 2)^2 = 13$ at the point $(3, 5)$.

2. Find the equation of the circle which has P$(2, -1)$ and Q$(8, 3)$ as opposite ends of a diameter.

3. Find the equation of the circle with centre $(1, 2)$ passing through the point $(-2, -1)$.

4. The straight line $y = x$ cuts the circle $x^2 + y^2 - 6x - 2y - 24 = 0$ at A and B. Find the coordinates of A and B.

5. Find the centre and radius of the circle $x^2 + y^2 - 2x - 2y - 3 = 0$.

6. Find the equation of the tangent to the circle in question 5 at the point A$(3, 2)$.

7. For what range of values of k does the equation $x^2 + y^2 - 6x + 4y + k = 0$ represent a circle.

8. Two circles with centres A and B and radii r and s $(r > s)$ touch. Express AB in terms of r and s, when the circles touch (i) externally (ii) internally.

SET 67

1. Write down the coordinates of the centre of the circle $2x^2 + 2y^2 - 6x + 9y - 2 = 0$.

2. The circle $x^2 + y^2 + 2gx + 2fy + c = 0$ passes through the origin. Write down the value of c.

3. The circle $x^2 + y^2 - 4x - 2y - 4 = 0$ is reflected in the x-axis. Find the equation of the image circle.

4. Find the equation of the circle on which A$(6, 7)$ and B$(2, -3)$ are diametrically opposite points.

5. State whether the point $(4, 3)$ lies **inside**, **outside** or **on** the circle $(x - 1)^2 + (y - 2)^2 = 25$.

6. Write down the centre and radius of the circle $x^2 + y^2 + 2x - 4y = 15$.

7. Find the gradient of the radius CA where C is the centre of the circle in question 6 and A is the point $(3, 4)$.

8. Find the equation of the tangent at A$(3, 4)$ to the circle in question 6.

SET 68

1. State the general equation of all circles which pass through the origin.
2. Write down the centre and radius of the circle with equation $x^2 + y^2 - 2x - 4y - 20 = 0$.
3. An identical circle to that in question 2 touches it externally at A(5, 5). State the coordinates of the centre of the other circle.
4. Find the equation of this second circle described in question 3.
5. The centre of a circle is (8, 12). The circle intersects the x-axis at (3, 0). Find the coordinates of the other point at which it intersects the x-axis.
6. Calculate the length of the tangent from the point A(11, −4) to the circle with equation $(x - 5)^2 + (y - 4)^2 = 51$.
7. Find the equation of the circle which passes through the points (0, 2) and (0, 8) and which touches the x-axis. [Two possible answers.]
8. Is the line with equation $y = 2x + 5$ a tangent to the circle $x^2 + y^2 - 2x - 4y = 0$?

ALL TOPICS IN UNIT 2

SET 69

1. The acute angles x and y are such that $\tan x = \frac{1}{3}$ and $\tan y = 1$. Evaluate $\sin (x + y)$ **exactly**.
2. Factorise fully $2x^3 + 4x^2 - 10x - 12$.
3. Find the equation of the tangent at (7, −1) to the circle $x^2 + y^2 - 6x + 8y = 0$.

4. ABCD,EFGH is a cuboid with AB = 2 cm, BC = 4 cm and AE = 3 cm. Calculate the angle between BH and the plane BCGF ($H\hat{B}C$).
5. Calculate the area of the shaded region.

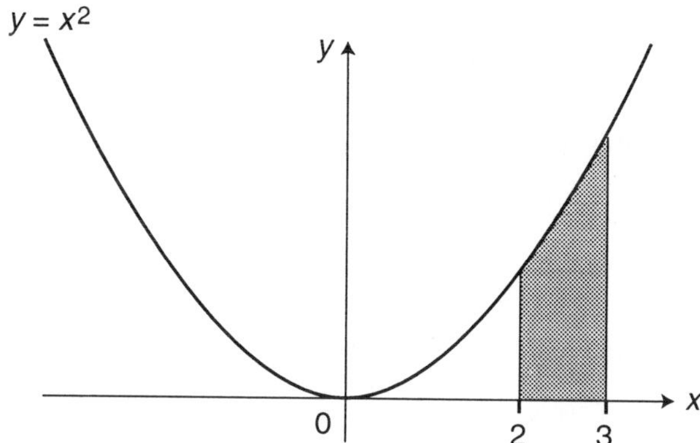

6. A curve for which $\frac{dy}{dx} = 15x^2 - 12x$ passes through the point (1, −1). Express y in terms of x.
7. Find the equation of the circle which passes through O(0, 0), A(6, 0) and B(0, 8).
8. Express $x^4 + 2x^2 + 5$ in the form $(x^2 + a)^2 + b$ and hence state its minimum value.

SET 70

1. Express cos A in terms of (i) $\cos \frac{A}{2}$ (ii) $\sin \frac{A}{2}$.

2. State in terms of θ and a the radius of the circle
$$x^2 + y^2 - 2ax \cos \theta - 2ay \sin \theta + a^2 \cos^2 \theta = 0.$$

3. Find the quotient and remainder when $x^3 - 4x + 10$ is divided by $x - 2$.

4. Calculate the size of the angle between any two faces of a regular tetrahedron.

5. Determine the nature of the roots of the equation $3x^2 - 4x - 2 = 0$.

6. A circle touches the y-axis at (0, 4) and the line $y = -1$ is also a tangent to the circle. Find the equation of the circle. [There are two possible answers.]

7. Evaluate $\int_{-1}^{3} (x^2 + 1)\, dx$.

8. Integrate $\frac{x^2 + 3x}{x}$ with respect to x.

SET 71

1. Find the exact solutions of $\sin x = \sin 2x$ for $0 \leq x \leq 2\pi$.

2. Determine the nature of the roots of the equation $x^2 - 3x = 28$.

3. Illustrate $\int_{-2}^{2} x^3\, dx$ by a sketch, and write down its value.

4. For acute angles x and y, $\sin x = \frac{12}{13}$ and $\sin y = \frac{3}{5}$. Show that the exact value of $\sin (x + y)$ is $\frac{63}{65}$.

5. Factorise fully $x^3 + 6x^2 + 13x + 10$.

6. Do the circles with equations $(x + 2)^2 + (y - 5)^2 = 1$ and $(x - 2)^2 + (y - 5)^2 = 9$ touch each other?

7. Integrate $\frac{x + 1}{\sqrt{x}}$ with respect to x.

8. Is the line $y = 2x + 5$ a tangent to the circle $x^2 + y^2 = 59$?

SET 72

1. Two concentric circles of radii 5 cm and 13 cm are drawn. A chord of the larger circle touches the inner circle. Calculate the length of the chord.

2. Factorise fully $2x^3 + 5x^2 - 23x + 10$.

3. ABCD,PQRS is a cube. Calculate the size of the angle between the planes ABCD and ABRS ($S\hat{A}D$).

4. By writing 105 = 60 + 45, find the **exact** surd value of sin 105°.

5. Write $(x + p)^2 = 3(x - q) - x^2$ as a quadratic equation in x in standard form.

6. Two identical circles touch at the point (5, 7). The equation of one is $(x - 3)^2 + (y - 2)^2 = 29$. Find the equation of the other one.

7. For a certain curve $\frac{dy}{dx} = 4x^3 - 6x$. Find the equation of the curve given that it passes through the point (2, –1).

8. Calculate the area between the curves $y = x^2$ and $y = 4x - x^2$.

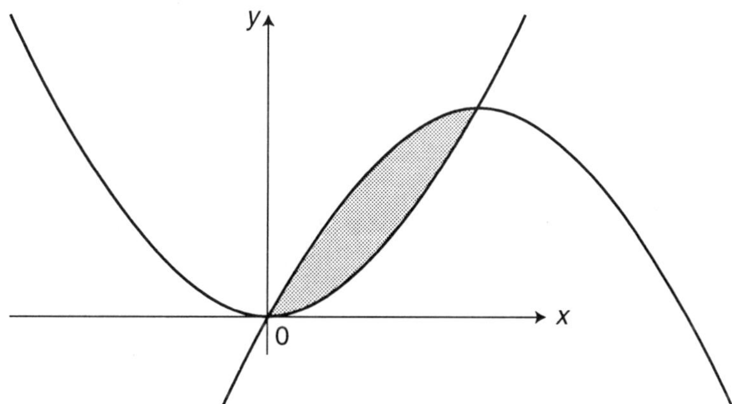

UNIT 3

REVISION OF UNITS 1 AND 2

SET 73

1. State the maximum and minimum values of the function f defined by $f(x) = 2(\cos x + 1)^2 - 3$.

2. For which values of k does the line $y = k$ not intersect the circle $x^2 + y^2 = 4$?

3. The line $y = mx + 3$ is perpendicular to the tangent at P(1, 0) to the curve $y = 1 - x^2$. Find m.

4. Evaluate $\int_1^4 x^{1/2}\,dx$.

5. Factorise fully (i) $2\sin^2 x + 5\sin x - 3$ (ii) $\cos^2 A + 2\sin A - 1$.

6. The sequence defined by $u_{n+1} = 0{\cdot}6u_n + r$ has a limit which is twice that of the sequence defined by $v_{n+1} = 0{\cdot}8v_n + s$. Express r in terms of s.

7. The graph of $y = f(x)$ is shown.

   Sketch the graph of $y = f'(x)$.

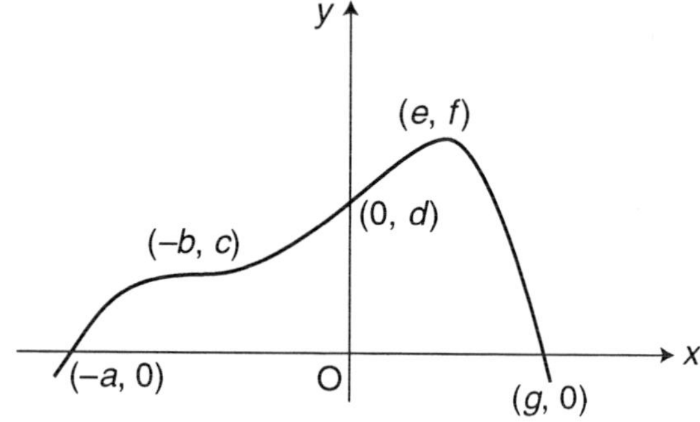

8. ΔABC has vertices A(3, 7), B(–2, –3) and C(8, 5). Find the equation of the median which passes through A.

SET 74

1. Write down the equation of the parabola $y = x^2$ after the translation of two units to the right.

2. Find the solution set of the inequality $x^2 - 5x - 24 < 0$, where x is a real number.

3. The line $3x + 2y = 13$ is the tangent at P to a circle with centre O. Find
   (i) the equation of OP (ii) the coordinates of P.

4. Find the equation of the circle in question 3.

5. $f: x \to \cos^2 x - \sin^2 x$. State the maximum value of f.

6. Find the rate of change of the function f given by $f(t) = 3t - t^2$ when $t = 1$.

7. ΔPQR has vertices P(0, 8), Q(–1, –3) and R(7, 5). Find the equation of the altitude from P.

8. Tadpoles in a pond are turning into frogs. One day at noon there were 250 frogs. Their number increased by 6% every day. How many days later was their number doubled?

SET 75

1. ABCD,EFGH is a cuboid with AB = 3 cm, BC = 8 cm and AE = 5 cm. M is the mid-point of FG. Calculate the size of the angle between AM and the base ABCD. ($M\hat{A}P$ where P is the mid-point of BC.)
2. The functions f and g are defined by $f(x) = 1 - 2x^2$, $g(x) = \sin x$. Express $f(g(x))$ in its simplest form.
3. Factorise $\cos^2 A - 3\cos A + 2$.
4. What is the size of the angle between the line $y = \sqrt{3}x + 2$ and the x-axis.
5. Write down the values of u_1 to u_4 for the sequence defined by $u_{n+1} = 3u_n - 4$ with $u_0 = 7$.
6. Differentiate $\dfrac{(x+1)^2}{\sqrt{x}}$ with respect to x.
7. Evaluate $\displaystyle\int_0^1 (\sqrt{x} + 1)^2\,dx$.
8. Find the coordinates of the points of intersection of the line $y = x - 2$ and the circle $x^2 + y^2 = 20$.

SET 76

1. Sketch the graph of $y = f(x)$ where $f: x \to x^2 - 6x + 9$.
2. If $\sin x° = \dfrac{1}{\sqrt{3}}$ $(0 < x < 90)$, find the exact values of (i) $\cos x°$ (ii) $\cos 2x°$.
3. Simplify $f(x) = \left(\sqrt[3]{x} - \dfrac{1}{\sqrt[3]{x}}\right)^2$. Find $f'(x)$.
4. What is the size of the angle between the line with equation $\sqrt{3}y = x - 4$ and the x-axis?
5. During the day bees remove 75% of the pollen from the stamens of an orchid. During the night the orchid produces 100 mg of pollen.
   Write down a recurrence relation to describe this situation.
6. Evaluate $\displaystyle\int (x^2 + 1)^2\,dx$.
7. Find the equation of the tangent to the circle $x^2 + y^2 = 29$ at the point $(-2, 5)$.
8. The graph of $f(x)$ is shown.

   Sketch the graph of
   (a) $-f(-x)$
   (b) $f(x + c)$

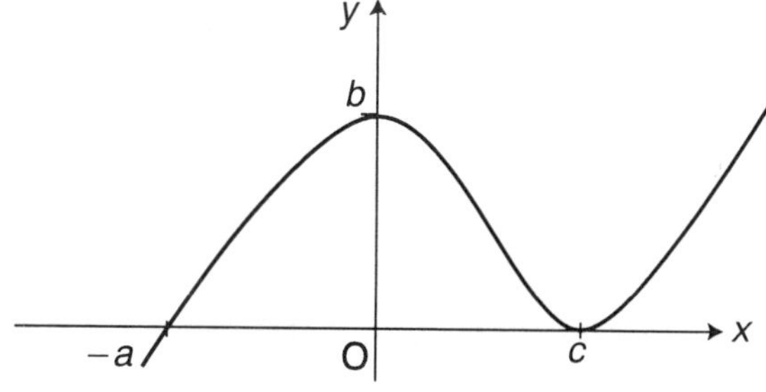

SET 77

1. Sketch the graph of $y = 10 - 3x - x^2$ stating the equation of the axis of symmetry.

2. Find the maximum value of (i) $\cos^2 x - \sin^2 x$ (ii) $2 \sin x \cos x$.

3. A is the point (6, 0) and B(0, 6). Write down the equation of the perpendicular bisector of AB.

4. Find the limit of the sequence generated by $u_{n+1} = 0{\cdot}3u_n + 14$.

5. Differentiate $\frac{1 + x}{x}$ with respect to x.

6. The line $y = k$ intersects the circle $(x - 2)^2 + (y - 3)^2 = 16$ in two points. Between which two values must k lie?

7. Given $f(x) = 2x + 3$ and $g(x) = x^2$, find a formula for $f(g(x))$.

8. Given that $f'(x) = 2x - 3$, and $f(3) = 4$, find $f(x)$.

SET 78

1. Write down the mid-point of AB where A is (5, –6) and B is (9, 8).

2. For a sector of a circle of radius 8 cm the angle at the centre is 30°. Calculate the length of the arc of the sector in terms of π.

3. Simplify $(1 - \sin^2 A) \tan^2 A$.

4. Solve $(x - 2)^2 - 16 \geq 0$.

5. By completing the square find the least possible value of $x^2 + 6x$.

6. Sketch the graph of $f(x) = x^2 + 5$ and hence state the range of the function f.

7. Find the equation of the tangent to $y = x^3 - 4x^2 + 11$ at the point (2, 3).

8. Find the area enclosed between $y = x^2 - 4$ and $y = -(x - 2)^2$.

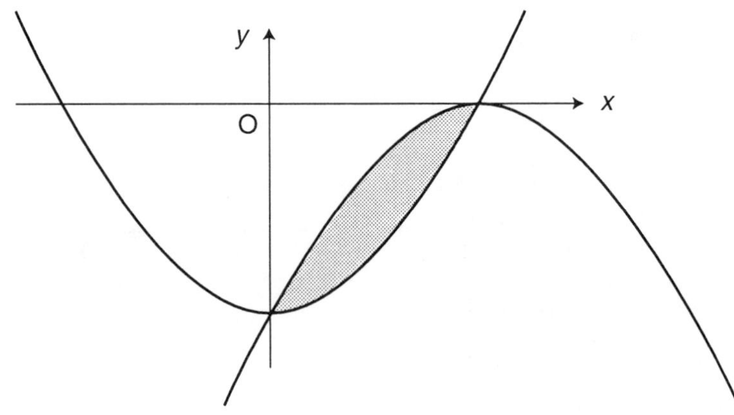

SET 79

1. What translation maps $y = x^2$ to $y = (x - 1)^2 + 2$?
2. Simplify $\sin^2 2A - \sin^2 A + \cos^2 2A - \cos^2 A$.
3. Arrange as a quadratic equation in x: $(x + c)^2 + x^2 = 9x$.
4. The lines $ay = 2x + 5$ and $y = 4x + 1$ are parallel. Find the value of a.
5. For what values of x is the curve with equation $y = x^3 - 3x^2 + 9$ stationary?
6. The sequences defined by $u_{n+1} = 0{\cdot}4u_n + a$ and $v_{n+1} = 0{\cdot}5v_n + a^2$ have the same limit. Find the value of a given that $a \neq 0$.
7. Calculate the area of the shaded region.

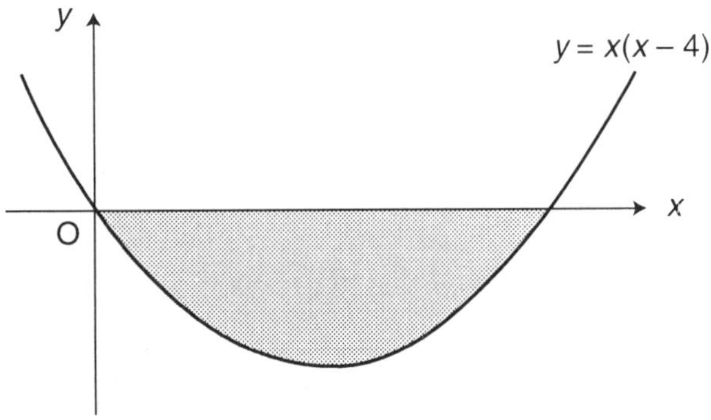

8. AB is a diameter of the smaller circle and a radius of the larger. The larger circle has equation $(x - 2)^2 + (y - 3)^2 = 36$. Find the equation of the smaller circle.

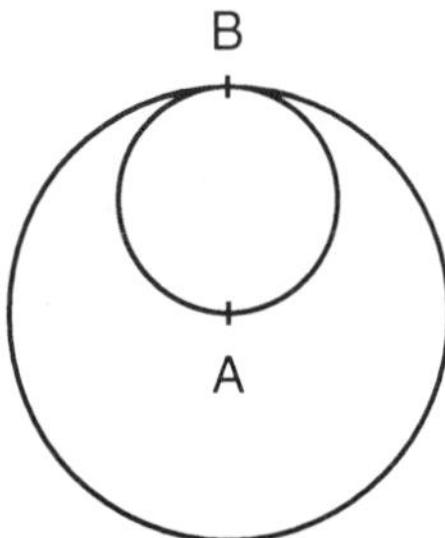

SET 80

1. For what value of x is there a stationary point on the curve $y = (1 - x)(x + 3)$?
2. $f(x) = x + 1$, $g(x) = 3x$ and $h(x) = x^2$. Evaluate $h\{g[f(2)]\}$.
3. Evaluate $\int_1^2 \frac{1}{x^2}\,dx$.
4. Solve $x(2 - 3x) - 2(3x + 5) \leq 4 - 3x(x - 1)$.
5. Simplify $\frac{1}{x-1} - \frac{x}{x^2 - 1}$
6. Arc PQ of a circle centre O radius 4 cm is 6 cm long. Calculate in radians the size of angle POQ.
7. Show that the lines $2x + 5y = 26$, $4x - y = 8$ and $3x + 4y = 25$ are concurrent, stating the coordinates of their point of intersection.
8. Find the equation of the circle which passes through the origin and touches the line $x = -2$ at the point $(-2, 4)$.

VECTORS

SET 81

1. Calculate the distance between
   (i) A(4, –7) and B(11,17) (ii) P(2, –1, 3) and Q(0, 2, –3).

2. If $\overrightarrow{PQ} = k\overrightarrow{QR}$ for some number k, what can be said about P, Q and R?

3. A is the point (0, 4), B(6, 0) and C(2, 2). M and N are the mid-points of AB and BC respectively. Write down the components of $\overrightarrow{MN}$.

4. A is (2, 3) and B is (–1, 6). P divides AB in the ratio 2 : 1. Calculate the coordinates of P.

5. Evaluate $\mathbf{a} \cdot \mathbf{b}$ where $\mathbf{a} = 2\mathbf{i} - 3\mathbf{j} + \mathbf{k}$ and $\mathbf{b} = 3\mathbf{i} - 2\mathbf{j} + 4\mathbf{k}$.

6. Simplify $\mathbf{x} \cdot (\mathbf{y} - \mathbf{z}) + \mathbf{y} \cdot (\mathbf{z} - \mathbf{x}) + \mathbf{z} \cdot (\mathbf{x} - \mathbf{y})$.

7. Calculate the cosine of the angle between the vectors $\begin{pmatrix} 2 \\ -1 \\ 2 \end{pmatrix}$ and $\begin{pmatrix} 3 \\ 6 \\ 2 \end{pmatrix}$.

8. V,ABCD is a square pyramid. $\overrightarrow{AB} = \mathbf{p}$, $\overrightarrow{AD} = \mathbf{q}$, $\overrightarrow{AV} = \mathbf{r}$. Express $\overrightarrow{VC}$ in terms of **p**, **q**, and **r**.

SET 82

1. The position vectors of P and Q are $\mathbf{p} = \begin{pmatrix} -2 \\ 1 \end{pmatrix}$ and $\mathbf{q} = \begin{pmatrix} 3 \\ -5 \end{pmatrix}$. Calculate $|\overrightarrow{PQ}|$.

2. $\overrightarrow{PQ} = \begin{pmatrix} 2 \\ 3 \end{pmatrix}$ and $\overrightarrow{QR} = \begin{pmatrix} a \\ b \end{pmatrix}$. If P, Q and R are collinear find a relation connecting a and b.

3. O(0, 0), P(6, –4) and Q(12, 5) are vertices of a quadrilateral OPQR.
   (i) Given that $\overrightarrow{OR} = \overrightarrow{PQ}$, find the coordinates of R.
   (ii) Show that OQ = PR.

4. R divides EF in the ratio 3 : 2. Find the coordinates of R if E is (5, 0, 10) and F is (0, 10, –5).

5. Find a if $2\mathbf{i} + 3\mathbf{j} - \mathbf{k}$ and $a\mathbf{i} + 2\mathbf{j} - 3\mathbf{k}$ are perpendicular.

6. Evaluate $(\mathbf{a} + \mathbf{b}) \cdot (\mathbf{a} - \mathbf{b})$.

7. Find the cosine of the angle between vectors $\mathbf{p} = 2\mathbf{i} + \mathbf{j} - 2\mathbf{k}$ and $\mathbf{q} = 3\mathbf{i} - 4\mathbf{k}$.

8. V,ABCD is a square pyramid. The diagonals of ABCD meet at E. M is the mid-point of VE. $\overrightarrow{AB} = \mathbf{p}$, $\overrightarrow{AD} = \mathbf{q}$, $\overrightarrow{AV} = \mathbf{r}$. Express $\overrightarrow{VM}$ in terms of **p**, **q** and **r**.

SET 83

1. Calculate the distance between the points (1, 5, 7) and (0, –3, 6).

2. X is the point (3, 4, 1), Y is (9, 1, –5) and Z is (11, 0, –7). Prove that X, Y and Z are collinear.

3. A and B are two points; M is the mid-point of AB. Write down an expression for
(i) $\overrightarrow{AB}$ in terms of **a** and **b** (ii) **m** in terms of **a** and **b**.

4. A is the point (–5, 7, –6) and B(5, 2, 9). Find the coordinates of C which divides AB in the ratio 3 : 2.

5. **i**, **j** and **k** are unit orthogonal vectors. Evaluate (i) **i** . **i** (ii) **j** . **k**.

6. Given that | **p** | = 3, | **q** | = 4 and **p** . **q** = 9, evaluate
(i) **p** . (**p** + **q**) (ii) (**p** – **q**) . **q**.

7. If **a** and **b** are vectors such that | **a** | = | **b** |, calculate the size of the angle between the vectors **a** + **b** and **a** – **b**.

8. $\mathbf{u} = \begin{pmatrix} 1 \\ 0 \\ 1 \end{pmatrix}$, $\mathbf{v} = \begin{pmatrix} 2 \\ -4 \\ 0 \end{pmatrix}$ and $3(\mathbf{x} + \mathbf{u}) = 2\mathbf{v} + \mathbf{x}$.
Calculate the components of **x**.

SET 84

1. Calculate the length of vector $4\mathbf{i} - 2\sqrt{3}\mathbf{j} + 2\sqrt{2}\mathbf{k}$.

2. For what value of *x* are the vectors $\begin{pmatrix} -2 \\ 3 \\ 6 \end{pmatrix}$ and $\begin{pmatrix} x \\ -9 \\ -18 \end{pmatrix}$ parallel?

3. $\mathbf{a} = \begin{pmatrix} 2 \\ 3 \end{pmatrix}$, $\mathbf{b} = \begin{pmatrix} 4 \\ 7 \end{pmatrix}$ and $\mathbf{c} = \begin{pmatrix} 8 \\ 11 \end{pmatrix}$.
Find *m* and *n* such that **c** = *m***a** + *n***b**.

4. P is the point (–3, 6, –9) and Q(7, 1, 6). Find the coordinates of R which divides PQ in the ratio 1 : 4.

5. Given that $\overrightarrow{AB} \cdot \overrightarrow{BC} = \frac{1}{2}$, evaluate (i) $\overrightarrow{BC} \cdot \overrightarrow{AB}$ (ii) $\overrightarrow{AB} \cdot \overrightarrow{CB}$.

6. The cosine of the angle between **p** and **q** is 0·25. |**p**| = 3 and |**q**| = 4. Evaluate **p** . (**p** + **q**).

7. Express $\cos P\hat{Q}R$ in terms of $\overrightarrow{QP}$ and $\overrightarrow{QR}$.

8. If $\mathbf{x} = \begin{pmatrix} 3 \\ 4 \end{pmatrix}$ and $\mathbf{y} = \begin{pmatrix} -5 \\ -12 \end{pmatrix}$ calculate $\dfrac{|\mathbf{x} + \mathbf{y}|}{|\mathbf{x}| + |\mathbf{y}|}$

SET 85

ABCD,EFGH is a cuboid with

$\overrightarrow{AB} = \begin{pmatrix} 4 \\ 8 \\ -4 \end{pmatrix}$, $\overrightarrow{AD} = \begin{pmatrix} 6 \\ -9 \\ -12 \end{pmatrix}$ and $\overrightarrow{AE} = \begin{pmatrix} 11 \\ -2 \\ 7 \end{pmatrix}$

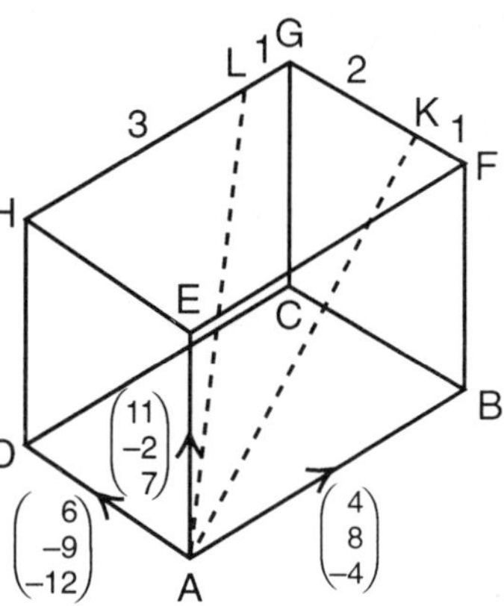

L divides HG in the ratio 3 : 1 and K divides FG in the ratio 1: 2.

1. Write down the components of $\overrightarrow{FK}$.
2. Write down the components of $\overrightarrow{HL}$ and $\overrightarrow{LG}$.
3. Express $\overrightarrow{AK}$ in terms of $\overrightarrow{AB}$, $\overrightarrow{AD}$ and $\overrightarrow{AE}$.
4. Hence find the components of $\overrightarrow{AK}$.
5. Similarly find the components of $\overrightarrow{AL}$.
6. Find the components of $\overrightarrow{LK}$.
7. Evaluate $\overrightarrow{LK} \cdot \overrightarrow{LG}$.
8. Show that $\cos G\hat{L}K = \dfrac{\sqrt{6}}{\sqrt{122}}$

SET 86

V,ABCD is a right rectangular based pyramid

with $\overrightarrow{AB} = \begin{pmatrix} 9 \\ 12 \\ 15 \end{pmatrix}$, $\overrightarrow{AD} = \begin{pmatrix} 5 \\ -10 \\ 5 \end{pmatrix}$ and $\overrightarrow{AV} = \begin{pmatrix} 14 \\ 2 \\ 5 \end{pmatrix}$

P divides DC in the ratio 1 : 2 and Q divides BC in the ratio 3 : 2.

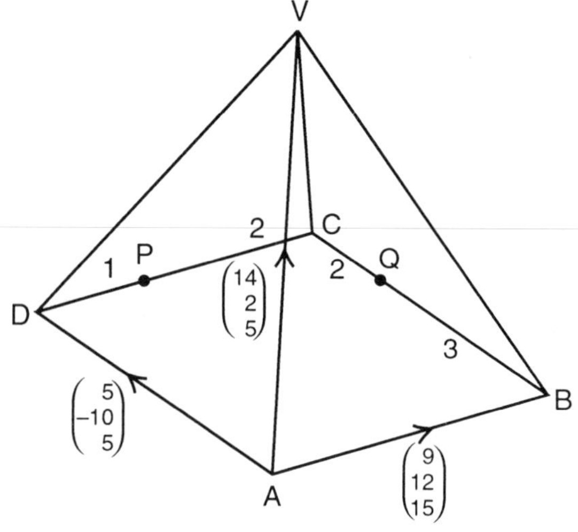

1. Express $\overrightarrow{AP}$ and $\overrightarrow{AQ}$ in terms of $\overrightarrow{AB}$ and $\overrightarrow{AD}$.
2. Hence find the components of $\overrightarrow{AP}$,
3. and the components of $\overrightarrow{AQ}$.
4. Evaluate $\overrightarrow{AP} \cdot \overrightarrow{AQ}$.
5. Express $\overrightarrow{VP}$ in terms of $\overrightarrow{AP}$ and $\overrightarrow{AV}$.
6. Hence find the components of $\overrightarrow{VP}$.
7. Similarly find the components of $\overrightarrow{VQ}$.

8. Calculate the size of $P\hat{V}Q$.

CALCULUS

SET 87

1. Differentiate $(x^2 + 4)^4$.

2. Differentiate $\frac{1}{1 - x}$

3. $f(x) = 2 + \sin x + \cos x$. Find $f'(x)$.

4. If $f(x) = \cos 3x$ find $f'\left(\frac{\pi}{2}\right)$

5. Evaluate $\int_{-\pi}^{\pi} (1 + \cos x)\, dx$.

6. Differentiate $\cos^2 x$.

7. Integrate with respect to x (i) $2x + 1$ (ii) $(2x + 1)^3$ (iii) $(2x + 1)^7$.

8. Calculate the area of the shaded region.

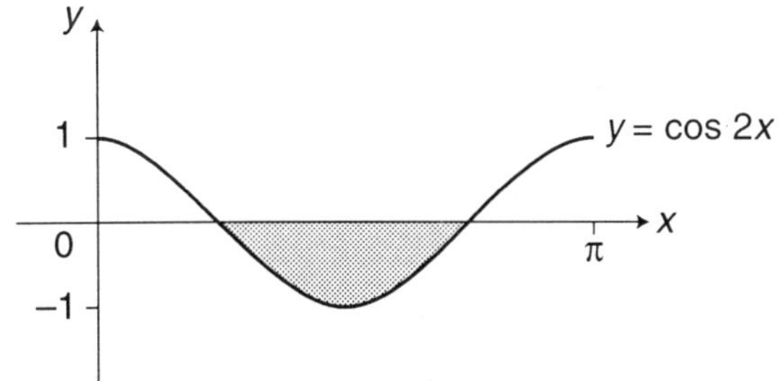

SET 88

1. If $f(x) = (2x^2 - 1)^3$, find $f'(x)$.

2. $f(x) = \frac{1}{\sqrt{x + 1}}$. Find $f'(x)$.

3. Find the derivative of $\cos (2x - 1)$.

4. Find the rate of change of the function f given by $f: x \to \sin x$ at the point $\left(\frac{3\pi}{2}, -1\right)$.

5. Calculate the area of the shaded region.

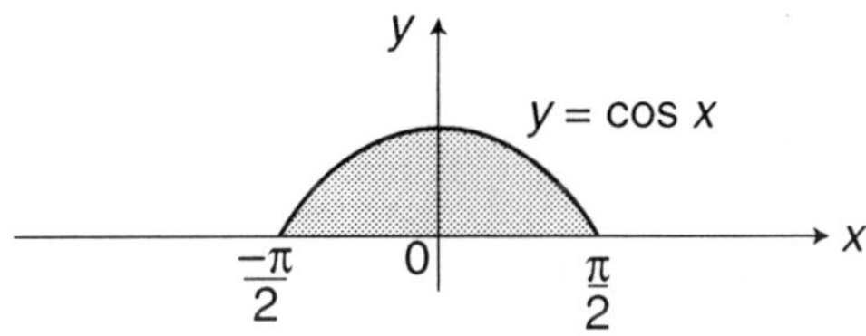

6. Find $f'(x)$ if $f(x) = (1 + \sin x)^4$.

7. Write down $\sin^2 u$ in terms of $\cos 2u$ and then find $\int \sin^2 u\, du$.

8. Calculate the area of the shaded region.

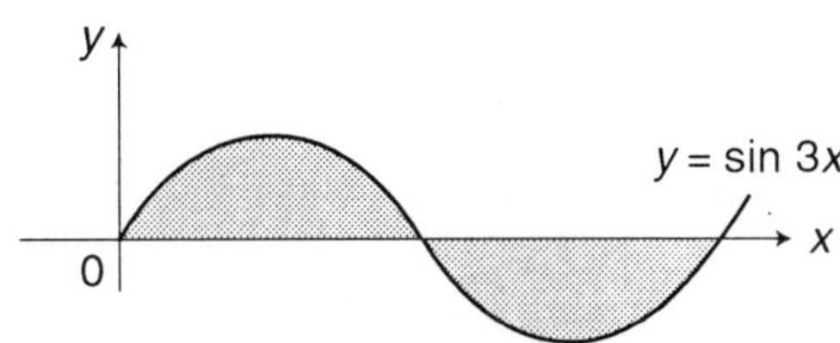

SET 89

1. If $y = (x^2 + 3x - 5)^4$ find $\frac{dy}{dx}$

2. Find $\frac{dy}{dx}$ if $y = \frac{2}{1 + x^2}$

3. Differentiate $\sin 2x$.

4. Find the gradient of the tangent to the curve $y = \cos 3x$ at the point $\left(\frac{\pi}{2}, 0\right)$.

5. Calculate the area of the shaded region.

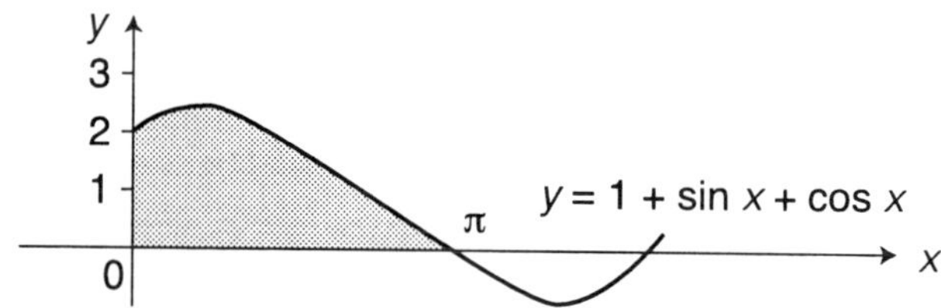

6. Differentiate $\sin^3 x$.

7. Integrate $\frac{1}{(3x + 4)^{2/3}}$ with respect to x.

8. Calculate the area of the shaded region.

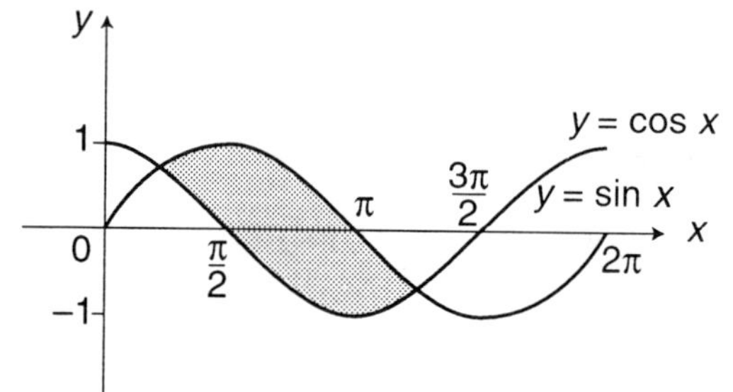

SET 90

1. Find $f'(x)$ if $f(x) = (x^3 + 5)^{3/2}$.

2. Find the derivative of $\sqrt{x^2 + 1}$.

3. Show that the function g defined by $g(x) = x + \sin x$ is never decreasing.

4. Calculate the gradient of the tangent to the curve $y = 2 \cos x$ at the point given by $x = \frac{\pi}{3}$.

5. Evaluate $\int_{-\pi/6}^{\pi/3} \cos x \, dx$

6. $f(x) = (1 - 2\cos x)^3$. Find $f'(x)$.

7. Evaluate $\int_3^4 (2x - 5)^3 \, dx$

8. Calculate the area of the shaded region.

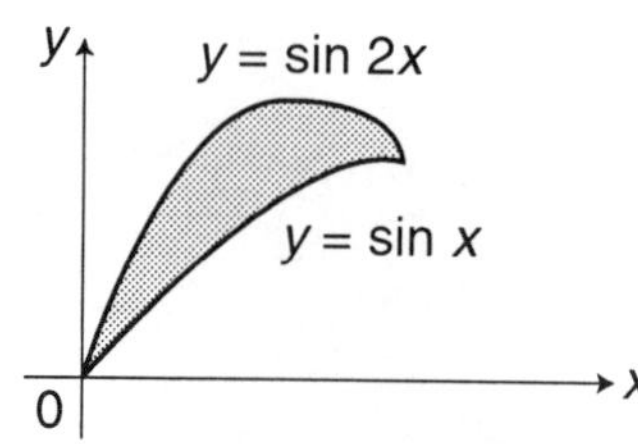

SET 91

1. Find the derivative of $(ax^2 + b)^n$.
2. Differentiate $\sqrt{(x+1)(x-1)}$.
3. $f(x) = \sin(x^2 + 2)$. Find $f'(x)$.
4. $f(x) = (1 - \sin x)(1 + \sin x)$. Find $f'\left(\frac{\pi}{4}\right)$.
5. Evaluate $\int_0^{\pi/2} (1 + \sin x)\, dx$
6. Differentiate $\cos^3 2x$.
7. Integrate with respect to α:
   (i) $\sin \alpha$ (ii) $\sin(2\alpha + 1)$ (iii) $\sin(2 - 3\alpha)$.
8. Calculate the area between $y = x^2$ and $y = 9$.

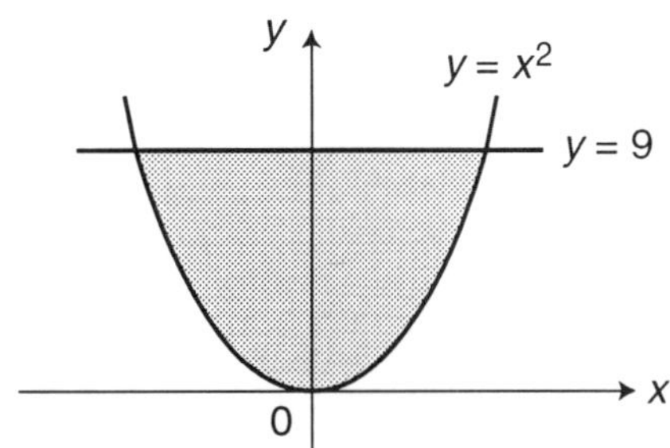

SET 92

1. Find the derivative of $(3x^2 - 5)^{2/3}$.
2. Find the derivative of $\frac{1}{\sqrt{(2x+3)(x-2)}}$.
3. Differentiate $\cos(3x^2 - 5)$.
4. Find $f'(-2)$ when $f(x) = \frac{1}{1 + x^2}$.
5. Evaluate $\int_{\pi/3}^{\pi/6} (\cos x - \sin x)\, dx$
6. $f: x \to 2x + 3$ and $g: x \to \cos x$ are functions on the set of real numbers. If $h(x) = g(f(x))$, find the derivative of h with respect to x.
7. Evaluate $\int_0^{\pi/2} (1 + \cos 2x)\, dx$
8. Calculate the area between $y = x^2$ and $y = 2x$.

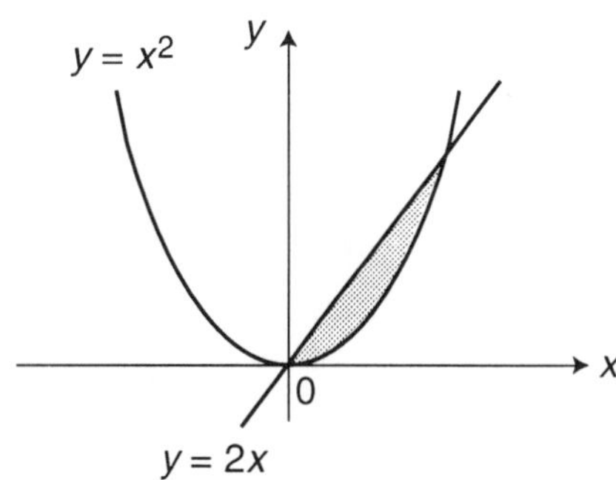

LOGARITHMS

SET 93

1. Express $y = e^x$ in logarithmic form.
2. Evaluate $\log_9 3$.
3. Solve $\log_2 x = 3$.
4. Express $3 + \log_3 t$ as a single logarithm.
5. Express $2\log_{10} 5 + \log_{10} 6$ in terms of $\log_{10} 1{\cdot}5$.

6. Solve $100 = 120.e^{-10k}$ for k.
7. Given that $R = 7S + 1{\cdot}58$, where $R = \log_{10} r$ and $S = \log_{10} s$, express r in terms of s.
8. Identify the logarithmic graph shown.

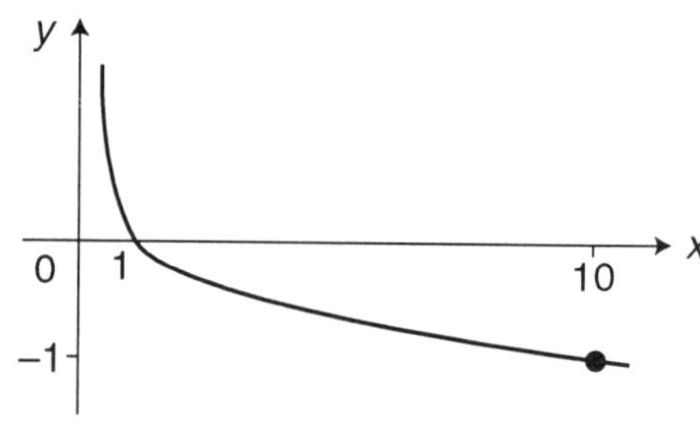

SET 94

1. Express $p = \log_e r$ in index form.
2. Evaluate $\log_4 8$.
3. Solve $x^{2{\cdot}5} = 32$.
4. Express $\log_2 8k - 2$ as a single logarithm.
5. Express $3\log_6 2 + \log_6 3$ in terms of $\log_6 4$.

6. Solve $50 = 80.e^{-30k}$ for k.

7. Given that $T = 2U + 1{\cdot}61$, where $T = \log_e t$ and $U = \log_e u$, express t in terms of u.
8. Identify the logarithmic graph shown.

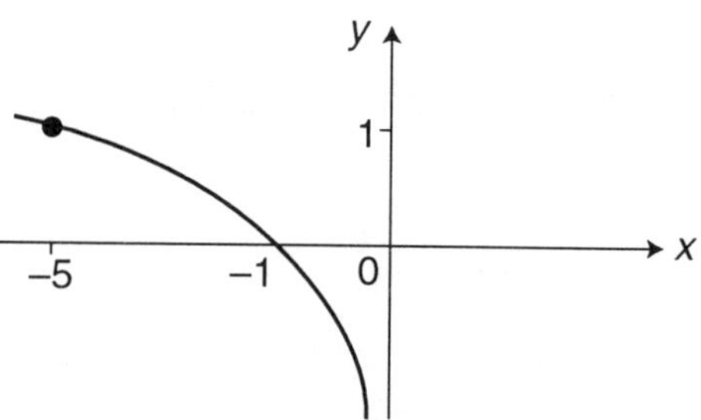

SET 95

1. Express $2^x = y$ in logarithmic form.
2. Evaluate $\log_4 64$.
3. Solve $\log_{\sqrt{3}} x = 8$
4. Express $1 + \log_3 2$ as a single logarithm.
5. Express $\frac{1}{2}\log_{10}(3^2 + 4^2)$ in terms of $\log_{10} 2$.
6. The atmospheric pressure at height h feet is given by $760.e^{-0{\cdot}002h}$ units. Calculate the atmospheric pressure at a height of 100 feet.
7. Given that $P = 4Q + 0{\cdot}6$, where $P = \log_{10} p$ and $Q = \log_{10} q$, express p in terms of q.
8. The graph of $y = \log_{10} x$ is shown by the full line. Write down the equation of the related graph shown by the dotted line

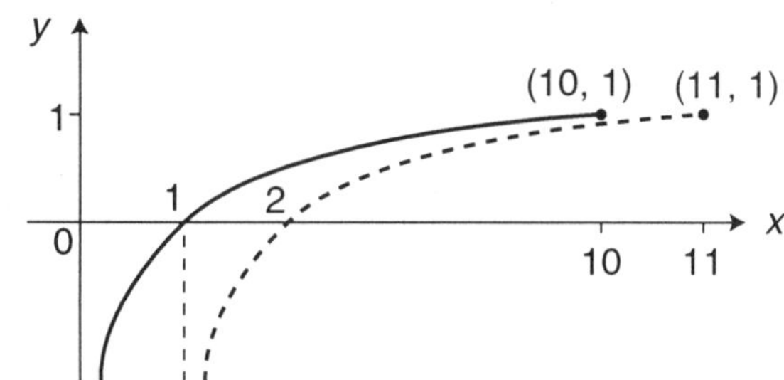

SET 96

1. Express in index form $p = \log_2 q$.
2. Evaluate $\frac{1}{2}\log_7 49$.

3. Solve $x^{2{\cdot}5} = 9$.
4. Express $3\log x - 4\log\sqrt{y}$ as a single logarithm.
5. Where does the graph of $y = \log_5 10x$ cross the x-axis?
6. The mass of a piece of radio-active material after t years is given by $500.e^{-0{\cdot}007\,t}$ mg. Calculate the mass after 10 years.
7. Given that $U = 1{\cdot}5V + 0{\cdot}4$, where $U = \log_{10} u$ and $V = \log_{10} v$, express u in terms of v.
8. Make a copy of the graph shown of $\log_5 x$ and superimpose the graph of $y = \log_5\left(\frac{1}{x}\right)$.

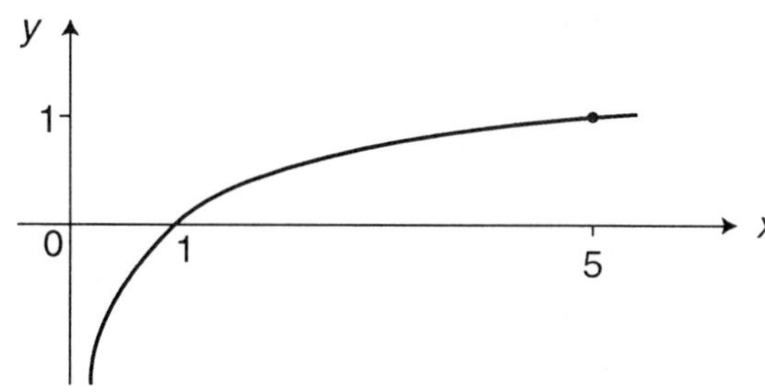

SET 97

1. Express $q^n = 3$ in logarithmic form.
2. Evaluate $\log_2 4\sqrt{2}$.

3. Solve $3^x = 7$.
4. Simplify $\log a + \log a^2 - \log a^3$.
5. Express log 54 in terms of log 4 and log 3. [Hint: $54 = 2 \times 27$]

6. The intensity of illumination, I_d, d metres through fog from a fog lamp of intensity I_0 is given by $I_d = I_0.e^{-0{\cdot}003d}$. At what distance is the intensity of illumination reduced by 10%?

7. Given that $Y = 6X + 2{\cdot}3$, where $Y = \log_e y$ and $X = \log_e x$, express y in terms of x.
8. The graph of $y = \log_3 x$ is shown by the full line. Write down the equation of the related graph, shown by the dotted line.

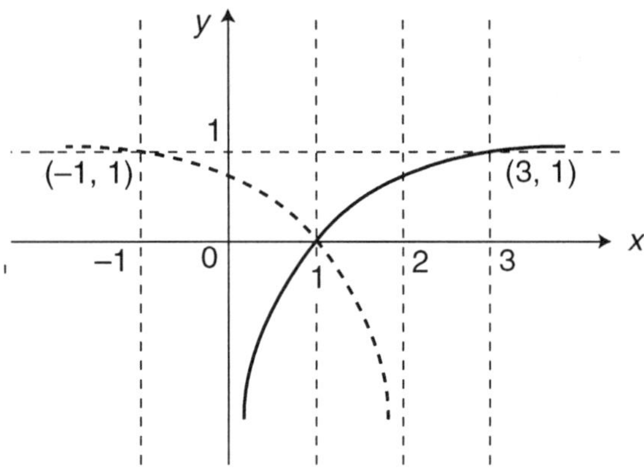

SET 98

1. Express $\log_a y = p$ as an index equation.
2. Evaluate $\log_5 125 + \log_8 64$.
3. Solve $\log_2 x = 5$.
4. Express $2 + \log_e 5$ as a single logarithm.
5. Where does the graph of $y = \log_{10} 2x$ cross the x-axis?

6. The mass, m_t, of a piece of radio-active element after t years decays according to the law $m_t = m_0.e^{-0{\cdot}007\,t}$. Calculate the half-life of the element.

7. Given that $Y = 0{\cdot}5 + 0{\cdot}3x$, where $Y = \log_{10} y$, express y in terms of x.
8. Make a copy of the graph shown of $y = \log_2 x$ and superimpose the graph of $y = \log_2 2x$.

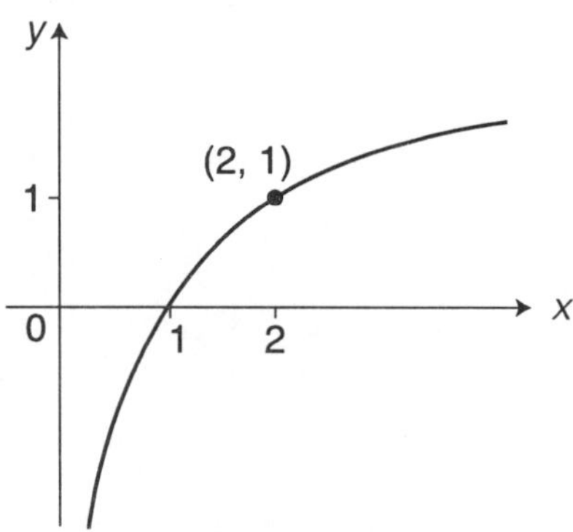

THE AUXILIARY ANGLE

SET 99

1. Write down the maximum value of $4\cos(x - 30)^\circ$ and state the value of x $(0 \leq x \leq 360)$ for which it occurs.
2. Write down the minimum value of $4\cos(x - 30)^\circ$ and state the value of x $(0 \leq x \leq 360)$ for which it occurs.

3. If $k\cos \alpha^\circ = 3$ and $k\sin \alpha^\circ = 4$, find the values of k and α, $k > 0$ and $0 \leq \alpha \leq 360$.
4. Express $2\cos x^\circ + 3\sin x^\circ$ in the form $k\cos (x - \alpha)^\circ$, $k > 0$ and $0 \leq \alpha \leq 360$.
5. Use your answer to question 4 to write down the maximum and minimum values of $2\cos x^\circ + 3\sin x^\circ$ and the values of x, $0 \leq x \leq 360$, for which they occur.
6. Find the maximum value of $\sqrt{3}\cos x^\circ + \sin x^\circ$, and the value of x, $0 \leq x \leq 360$, for which it occurs.
7. Use your working for question 6 to sketch the graph of $y = \sqrt{3}\cos x^\circ + \sin x^\circ$ for $0 \leq x \leq 360$, showing where the curve crosses the coordinate axes.

8. Solve $4\cos x^\circ - 3\sin x^\circ = 1$ in the interval $0 \leq x \leq 360$.

SET 100

1. Write down the maximum value of $7\sin(x + 40)^\circ$ and state the value of x $(0 \leq x \leq 360)$ for which it occurs.
2. Write down the minimum value of $7\sin(x + 40)^\circ$ and state the value of x $(0 \leq x \leq 360)$ for which it occurs.

3. If $k\cos \alpha^\circ = 5$ and $k\sin \alpha^\circ = -12$, find the values of k and α, $k > 0$ and $0 \leq \alpha \leq 360$.
4. Express $2\sin x^\circ + \sqrt{5}\cos x^\circ$ in the form $k\sin (x + \alpha)^\circ$, $k > 0$ and $0 \leq \alpha \leq 360$.
5. Use your answer to question 4 to write down the maximum and minimum values of $2\sin x^\circ + \sqrt{5}\cos x^\circ$ and the values of x, $0 \leq x \leq 360$, for which they occur.

6. Find the minimum value of $2\cos x^\circ - 3\sin x^\circ$, and the value of x, $0 \leq x \leq 360$, for which it occurs.
7. Use your working for question 6 to sketch the graph of $y = 2\cos x^\circ - 3\sin x^\circ$ for $0 \leq x \leq 360$, showing where the curve crosses the coordinate axes.

8. Solve $6\sin x^\circ + 8\cos x^\circ = 7$ in the interval $0 \leq x \leq 360$.

SET 101

1. Write down the maximum value of $2 + 3\cos(x + 20)^\circ$ and state the value of x $(0 \leq x \leq 360)$ for which it occurs.
2. Write down the minimum value of $2 + 3\cos(x + 20)^\circ$ and state the value of x $(0 \leq x \leq 360)$ for which it occurs.
3. If $k\cos \alpha^\circ = -1$ and $k\sin \alpha^\circ = 1$, find the values of k and α, $k > 0$ and $0 \leq \alpha \leq 360$.

4. Express $\cos x° + \sin x°$ in the form $k\sin(x + \alpha)°$, $k > 0$ and $0 \le \alpha \le 360$.

5. Use your answer to question 4 to write down the maximum and minimum values of $\cos x° + \sin x°$ and the values of x, $0 \le x \le 360$, for which they occur.

6. Find the minimum value of $3\cos x° + 4\sin x°$, and the value of x, $0 \le x \le 360$, for which it occurs.

7. Use your working for question 6 to sketch the graph of $y = 3\cos x° + 4\sin x°$ for $0 \le x \le 360$, showing where the curve crosses the coordinate axes.

8. Solve $2\cos x + 3\sin x = 1$ in the interval $0 \le x \le 2\pi$. [Notice the use of radians.]

SET 102

1. Write down the maximum value of $5 - \sin(2x + 60)°$ and state the value of x ($0 \le x \le 180$) for which it occurs.

2. Write down the minimum value of $5 - \sin(2x + 60)°$ and state the value of x ($0 \le x \le 180$) for which it occurs.

3. If $k\cos \alpha° = -\sqrt{2}$ and $k\sin \alpha° = -\sqrt{2}$, find the values of k and α, $k > 0$ and $0 \le \alpha \le 360$.

4. Express $4\cos x° - 3\sin x°$ in the form $k\cos(x + \alpha)°$, $k > 0$ and $0 \le \alpha \le 360$.

5. Use your answer to question 4 to write down the maximum and minimum values of $4\cos x° - 3\sin x°$ and the values of x, $0 \le x \le 360$, for which they occur.

6. Find the minimum value of $\cos x - \sin x$, and the value of x, $0 \le x \le 2\pi$, for which it occurs. [Notice that x is measured in radians.]

7. Use your working for question 6 to sketch the graph of $y = \cos x - \sin x$ for $0 \le x \le 2\pi$, showing where the curve crosses the coordinate axes.

8. Solve $5\sin x - 2\cos x = 3$ in the interval $0 \le x \le 2\pi$.

SET 103

1. Write down the maximum value of $1 + 2\sin\left(x + \frac{\pi}{3}\right)$ and state the value of x ($0 \le x \le 2\pi$) for which it occurs.

2. Write down the minimum value of $1 + 2\sin\left(x + \frac{\pi}{3}\right)$ and state the value of x ($0 \le x \le 2\pi$) for which it occurs.

3. If $k\cos \alpha° = 1$ and $k\sin \alpha° = -1$, find the values of k and α, $k > 0$ and $0 \le \alpha \le 360$.

4. Express $3\cos x° + 2\sin x°$ in the form $k\cos(x - \alpha)°$, $k > 0$ and $0 \le \alpha \le 360$.

5. Use your answer to question 4 to write down the maximum and minimum values of $3\cos x° + 2\sin x°$ and the values of x, $0 \le x \le 360$, for which they occur.

6. Find the maximum value of $\cos x + \sqrt{3}\sin x$, and the value of x, $0 \le x \le 2\pi$, for which it occurs.

7. Use your working for question 6 to sketch the graph of $y = \cos x + \sqrt{3}\sin x$ for $0 \le x \le 2\pi$, showing where the curve crosses the coordinate axes.

8. Solve $2\cos x° - 3\sin x° = 2{\cdot}5$ in the interval $0 \le x \le 360$.

SET 104

1. Write down the maximum value of $2 - 3\cos\left(x - \frac{\pi}{4}\right)$ and state the value of x $(0 \le x \le 2\pi)$ for which it occurs.
2. Write down the minimum value of $2 - 3\cos\left(x - \frac{\pi}{4}\right)$ and state the value of x $(0 \le x \le 2\pi)$ for which it occurs.

3. If $k\cos\alpha = -2$ and $k\sin\alpha = -3$, find the values of k and α, $k > 0$ and $0 \le \alpha \le 2\pi$.
4. Express $3\cos x° - 5\sin x°$ in the form $k\cos(x + \alpha)°$, $k > 0$ and $0 \le \alpha \le 360$.
5. Use your answer to question 4 to write down the maximum and minimum values of $3\cos x° - 5\sin x°$ and the values of x, $0 \le x \le 360$, for which they occur.
6. Find the maximum value of $\sin x - \cos x$, and the value of x, $0 \le x \le 2\pi$, for which it occurs.
7. Use your working for question 6 to sketch the graph of $y = \sin x - \cos x$ for $0 \le x \le 2\pi$, showing where the curve crosses the coordinate axes.

8. Solve $\sqrt{2}\sin x° + \sqrt{3}\cos x° = \sqrt{5}$ in the interval $0 \le x \le 360$.

ALL TOPICS IN UNIT 3

SET 105

1. A is the point (3, –1, 2) and B is (2, 5, –3). Evaluate **a . b**.

2. Express $2\cos x° - \sin x°$ in the form $k\cos(x + \alpha)$, $k > 0$ and $0 \le \alpha \le 360$.
3. Express log 15 + log 10 – 2 log 5 as the log of a single number.
4. Evaluate $\int_{-1}^{3} \sqrt{1 + t}\, dt$.
5. Draw a counter-example to disprove the conjecture **a . b** = **b . c** ⇒ **a** = **c**.
6. $f: x \to \cos x + \sin x$. State the period of f.
7. The graph shown has equation $y = \log_b x$.
The point (3, 1) lies on this graph. Find the value of b.
The point (a, 2) lies on this graph. Find the value of a.

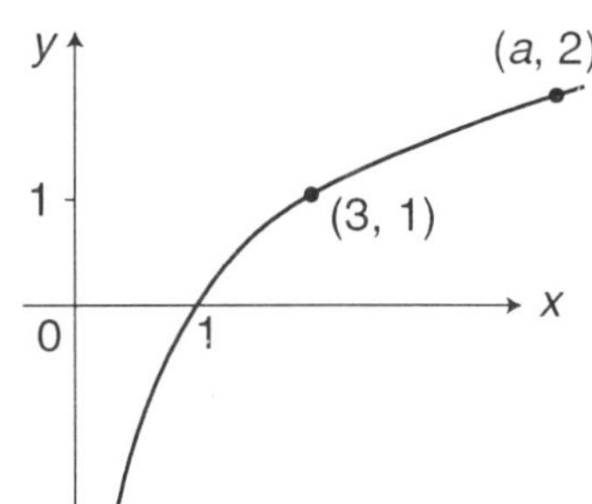

8. Differentiate $\dfrac{1}{\sqrt{1 - x}}$.

SET 106

1. If $\overrightarrow{AB} = \overrightarrow{DC}$, what kind of figure is ABCD.
2. State the maximum value of $5\cos(x - 40)°$ and the value of x $(0 \le x \le 360)$ for which it occurs.
3. Express $2 + \log_3 5$ as the logarithm of a single number.

4. Integrate $\dfrac{1}{2(x-2)^2}$

5. If $\overrightarrow{PR} \cdot \overrightarrow{PQ} = -1$, what can be said about the size of $R\hat{P}Q$?

6. Find the maximum value of $1 + 3\cos x - 4\sin x$.

7. The graph shown has equation
$y = \log_b (x + a)$.
Find the values of a and b.

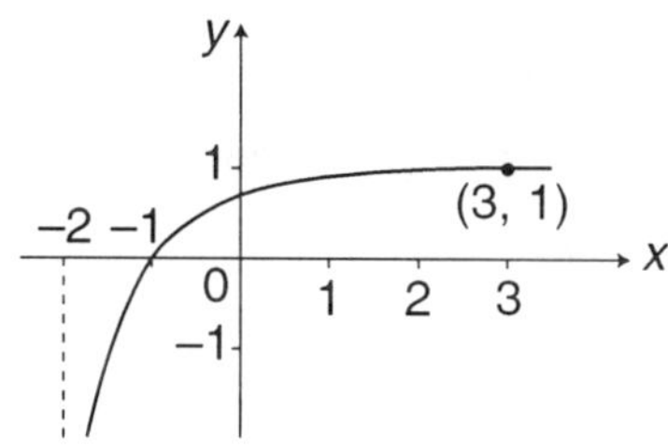

8. Find the derivative of $\sin (2x^2 + 1)$.

SET 107

1. Find the coordinates of C which divides AB in the ratio 3 : 2, where A is (–2, 10, –10) and B(8, –5, 10).

2. Express $3\cos x° + 5\sin x°$ in the form $k\cos (x - \alpha)°$, $k > 0$ and $0 \le \alpha \le 360$.

3. Evaluate $\log_9 \sqrt{3}$.

4. Integrate $\cos (2x + 1)$ with respect to x.

5. The three sides, each of length 2 units, of an equilateral triangle are represented by **a**, **b** and **c** as shown.
Evaluate **a** . (**a** + **b**).

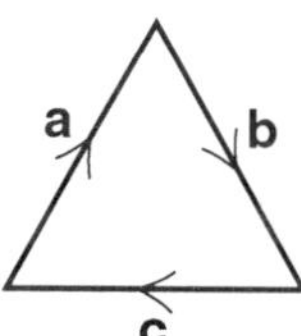

6. Find the minimum value of $4 - 5\cos x + 12\sin x$.

7. Given that $P = 3Q + 1{\cdot}04$, where $P = \log_{10} p$ and $Q = \log_{10} q$, express p in terms of q.

8. Find the derivative of $\cos (5x + 4)$.

SET 108

1. Find the coordinates of R which lies on PQ (produced) such that $3\overrightarrow{PR} = 4\overrightarrow{PQ}$, where P is (–4, 7, 1) and Q is (5, 1, 4).

2. Write down the minimum value of $3 + \sqrt{2}\cos (x + 20)°$ and the value of x ($0 \le x \le 360$) for which it occurs.

3. Express $\log \dfrac{24}{27}$ in terms of log 2 and log 3.

4. Integrate $\dfrac{2}{(5x+3)^4}$ with respect to x.

5. The angle between **a** and **b** is 45°
$|\mathbf{a}| = 3$ and $|\mathbf{b}| = \sqrt{2}$.
Evaluate **a.a**, **b.b**, **a.b** and hence **p.p** where **p** = 2**a** + 3**b**.

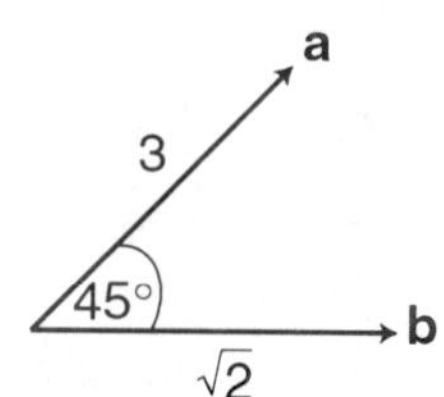

6. Express $3\cos x° + 2\sin x°$ in the form $k\sin(x + \alpha)°$, $k > 0$ and $0 \le \alpha \le 360$.

7. Given that $s = 5e^{4t}$, express $\log_e s$ in terms of t.

8. Write down the derivative of $(x^2 - 5x - 2)^{3/2}$.

WHOLE COURSE REVISION

SET 109

State whether each of the following is **true** or **false**:

1. $(a + b)^{-n} = a^{-n} + b^{-n}$

2. $2^3 \times 3^4 = 6^7$

3. $\log (X + Y) = \log X + \log Y$

4. $\int \sin^2 x\, dx = \dfrac{\sin^3 x}{3} + c$

5. $\dfrac{\log_a M}{\log_a N} = \log_a \dfrac{M}{N}$

6. $\sin 4x = 4\sin x \cos x$

7. If $f(x) = 3x^2 + 7$ then $f'(x) = 6x + 7$.

8. $\int_a^b f(x)\, dx = -\int_b^a f(x)\, dx$

SET 110

1. Sketch without calculation the graphs of
(i) $y = x^2 + 1$ (ii) $y = (x + 1)^2$.

2. Write down the equation of the line through P(h, k) parallel to the y-axis.

3. State the condition for $x - p$ to be a factor of $f(x)$.

4. The point (h, k) lies inside the circle $x^2 + y^2 = r^2$. State a relationship connecting h, k and r.

5. A is (a_1, a_2, a_3) and B is (b_1, b_2, b_3). Write down the component form of **a . b**.

6. Simplify (i) $\sin (90° - A)$ (ii) $\sin (180° - A)$.

7. The graph shown has equation $y = a\cos (bx + c)$.
State the values of a, b and c.

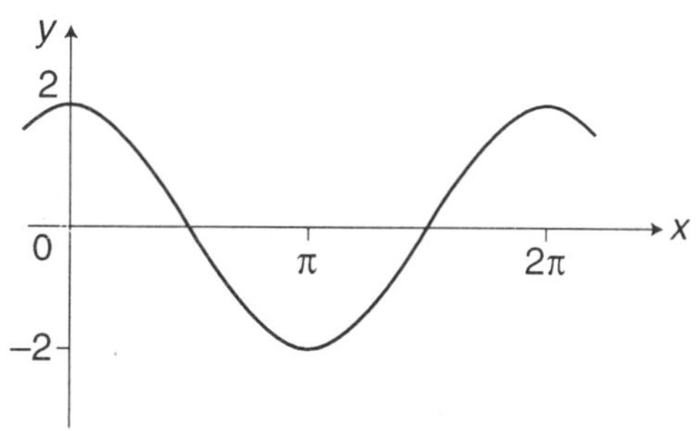

8. Sketch the curve $y = \sin 2x$ for $0 \le x \le 2\pi$ and shade the area represented by $\int_0^{\pi/2} \sin 2x\, dx$

SET 111

1. Simplify (i) $(2x)^{-1}$ (ii) $2x^{-1}$

2. Simplify (i) $1 - \cos^2 A$ (ii) $1 - \sin^2 A$.

3. Given that $\log_a b = x$, express b in terms of a and x.

4. A function f defined on the real numbers is given by $f(x) = \dfrac{ax^2 + bx + c}{(x-k)(x-l)}$.
State any restrictions which must be placed on x.

5. If $\overrightarrow{OB} - \overrightarrow{OA} = \overrightarrow{OC} - \overrightarrow{OD}$, what kind of figure is ABCD.

6. Write down an expression for the distance between A(x_1, y_1, z_1) and B(x_2, y_2, z_2).

7. Write down an expression for:
(i) $\cos^2 A$ in terms of $\cos 2A$ (ii) $\sin^2 A$ in terms of $\cos 2A$.

8. If $p + q = 2$ and $pq = 1$, find the value of $\dfrac{1}{p} + \dfrac{1}{q}$.

SET 112

1. When does the sequence defined by the recurrence relation $u_{n+1} = au_n + b$ have a limit as n tends to infinity?

2. State the condition for the equation $px^2 + qx + r = 0$ to have no real roots.

3. Write down an expression for the length of the vector $\begin{pmatrix} a \\ b \\ c \end{pmatrix}$.

4. R is the point (r_1, r_2, r_3) and S is (s_1, s_2, s_3).
(i) Give the coordinates of T, the mid-point of RS.
(ii) Write down **r . s** in component form.

5. Given that $225° = k\pi$ radians, k a real number, find the value of k.

6. (i) The graph shown has equation $y = a\sin(bx + c)$.
State the values of a, b and c.
(ii) State the period of function f.

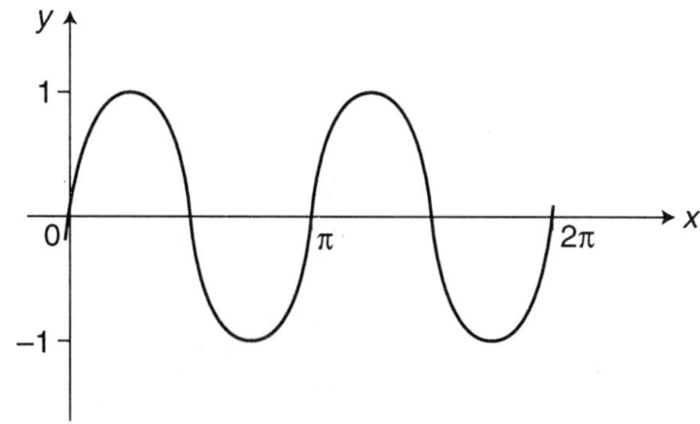

7. Differentiate $\sin^2 5x$.

8. Integrate $\sin^2 5x$.

SET 113

1. Give the maximum value of (i) $\sin x$ (ii) $\sin 2x$ (iii) $\sin \frac{x}{2}$.
2. Complete $a^x = b \Rightarrow \log \ldots$
3. A circle passes through the points P(a, 0), Q(0, b) and the origin. State the coordinates of the centre of the circle.
4. If $\overrightarrow{AB} = t\overrightarrow{BC}$ (t a constant), what can be said about A, B and C?
5. Simplify
(i) $\sin^2 3A + \cos^2 3A$ (ii) $\sin^2 A - \cos^2 A$ (iii) $2 \cos \theta \sin \theta$.
6. Sketch the graph of the function $\cos 2x$, $0 \leq x \leq 2\pi$.
7. State $\frac{d}{dx}(\cos x) + \frac{d}{dx}(\sin x)$.
8. If $\int_a^b f(x)\,dx = k$ and $\int_b^c f(x)\,dx = l$, find $\int_a^c f(x)\,dx$.

SET 114

1. Sketch, without calculation, the graph of
(i) $y = x^2$ (ii) $y = x^3$.
2. Evaluate (i) a^0 (ii) $\log_a a$ (iii) $\log_a 1$.
3. **i**, **j** and **k** are unit orthogonal vectors. Evaluate **k . k** and **k . i** .
4. A is (x_1, y_1) and B is (x_2, y_2). The line AB makes an angle θ with OX. Write down an expression for $\tan \theta$.
5. Simplify
(i) $\cos A \cos B - \sin A \sin B$ (ii) $\cos X \sin Y - \sin X \cos Y$.
6. State the maximum value of
(i) $2 \cos x$ (ii) $3 \sin (x - 20)°$ (iii) $\frac{1}{2 - \cos x}$.
7. State an expression for the gradient of the curve $y = f(x)$ at point $(a, f(a))$.
8. The graph of the function f shown has half-turn symmetry about the origin.
Evaluate $\int_{-1}^{1} f(x)\,dx$.

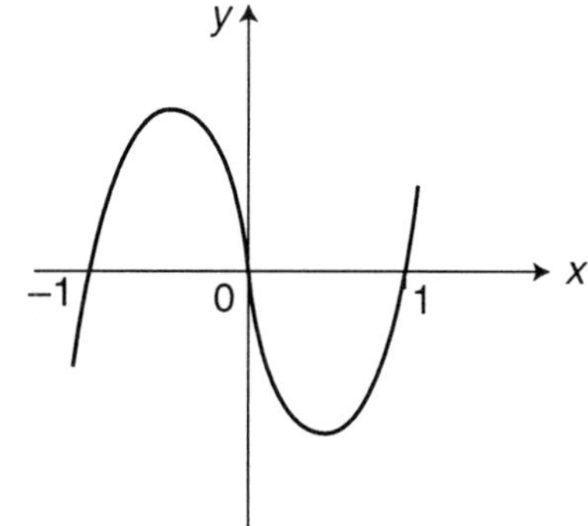

SET 115

1. Write down the "quadratic formula" for the solution of the equation $hx^2 + kx + l = 0$.
2. A is (x_a, y_a) and B is (x_b, y_b). State the coordinates of the mid-point of AB.
3. Copy and complete
   (i) $a^{-n} = \ldots$ (ii) $a^{1/n} = \ldots$ (iii) $\log_a X - \log_a Y = \ldots$ (iv) $\log_b XY = \ldots$
4. State the equation of the straight line through (h, k) with gradient n.
5. The point P(**p**) divides MN in the ratio 2 : 1. Express **p** in terms of **m** and **n**.
6. State the period of
   (i) $\tan x$ (ii) $\tan 2x$ (iii) $\sin 3x$ (iv) $\cos \frac{x}{2}$ (v) $\sin \frac{x}{4}$.
7. If $f(x) = \sin x$, state (i) $f'(x)$ (ii) $\int f(x)dx$.
8. Write down an expression for the area enclosed by the curve $y = g(x)$ shown, the x-axis and lines $x = a$ and $x = b$.

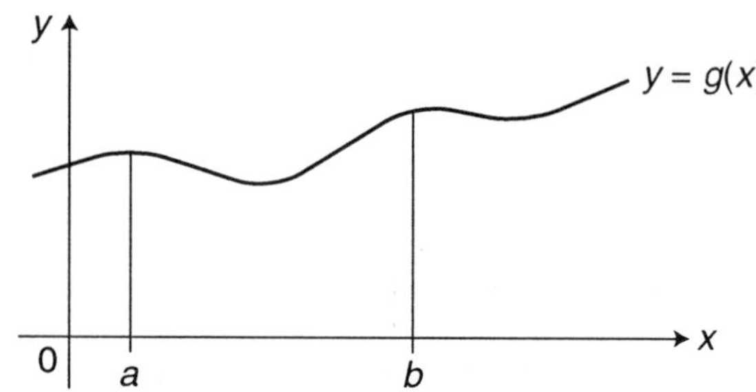

SET 116

1. The graph of the function f as defined by $f(x) = ax^2 + bx + c$ crosses the x-axis at points A(–2, 0) and B(4, 0). State the equation of the axis of symmetry of the graph.
2. Express the equation $y = ax^n$ in equivalent logarithmic form.
3. Rationalise the denominators in the expressions (i) $\frac{1}{\sqrt{a}}$ (ii) $\frac{1}{\sqrt{a} - b}$.
4. A radius OP of a circle, centre O, has gradient m. State the gradient of the tangent at P.
5. Sketch the graph of $\sin 2x°$, $0 \leq x \leq 360$.
6. If $k\cos \alpha = 4$ and $k\sin \alpha = -3$, state the value of k, given that $k > 0$.
7. Describe $f'(x)$ when $f(x)$ is increasing.
8. Evaluate $\int \sin 2x \, dx$.

SET 117

1. Evaluate $\dfrac{\log_5 32}{\log_5 8}$
2. The expression $ax^2 + bx + c$ is divided by $x - h$. State an expression for the remainder.
3. The circle $x^2 + y^2 + 2gx + 2fy + c = 0$ passes through the origin. State the value of c.
4. Two vectors **a** and **b** are at right angles. State the value of **a . b** .
5. Express
   (i) cos 2A in terms of cos A (ii) cos 4A in terms of cos 2A
   (iii) cos A in terms of $\sin \dfrac{A}{2}$.
6. Write down the maximum value of
   (i) $\cos x$ (ii) $2\cos x$ (iii) $-2\cos x$.
7. State an expression for the rate of change of the function $y = f(x)$ when $x = b$.
8. Evaluate $\int_0^2 k\,dx$, k constant.

SET 118

1. The graph of the function $f: x \to a^x$ cuts the y-axis at the point A. State the coordinates of A.
2. State a formula involving $\hat{P}$ for the area of ΔPQR.
3. Simplify $\log_a X + \log_a Y + \log_a Z$.
4. The centre of a circle is (u, v) and the radius is r. Write down the equation of the circle.
5. Simplify
   (i) $\cos^2 \frac{x}{3} + \sin^2 \frac{x}{3}$ (ii) $\cos^2 \frac{x}{2} - \sin^2 \frac{x}{2}$ (iii) $2\sin \frac{x}{2} \cos \frac{x}{2}$.
6. State the period of (i) $\cos x$ (ii) $\cos 2x$ (iii) $\cos 3x$.
7. $f(x) = (ax + b)^n$. Write down $f'(x)$.
8. Simplify $\int_a^b f(x)\,dx + \int_b^c f(x)\,dx$.

SET 119

1. If $f(x) = x^3 - 6x^2 + 8x$, write down, from the graph, an expression for the related function $g(x)$.

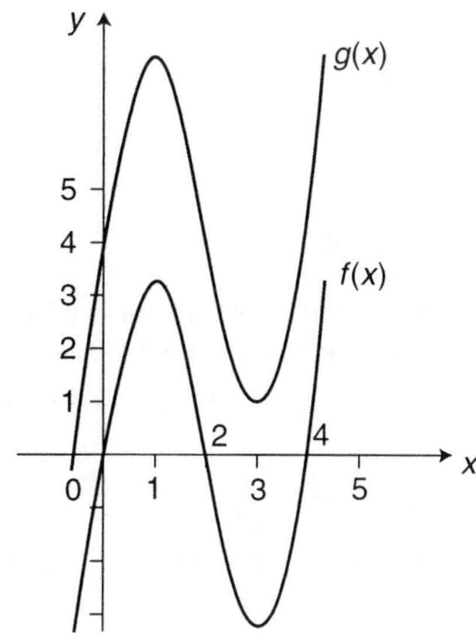

2. Find the cubic expression for $f(x)$, shown in this graph.

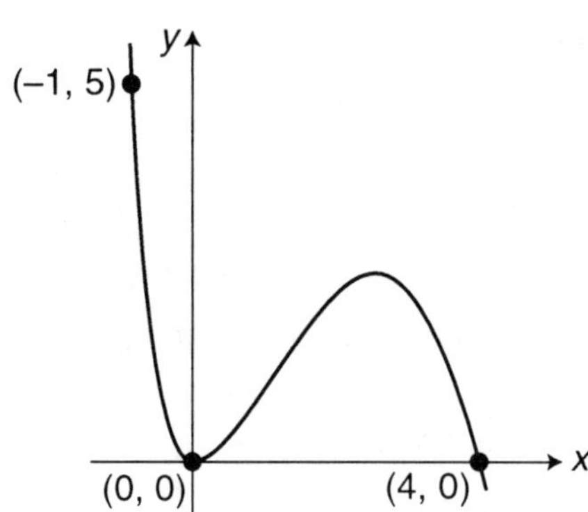

3. (i) AM is a median of triangle ABC, describe AM.
(ii) BD is an altitude of triangle ABC, describe BD.

4. **a** and **b** are vectors where **a . b** = 2. Evaluate (i) **b . a** (ii) 3**a . b**.

5. The diagram shows the graph of the function f defined by $f(x) = k + \sin x$.
State the value of k.

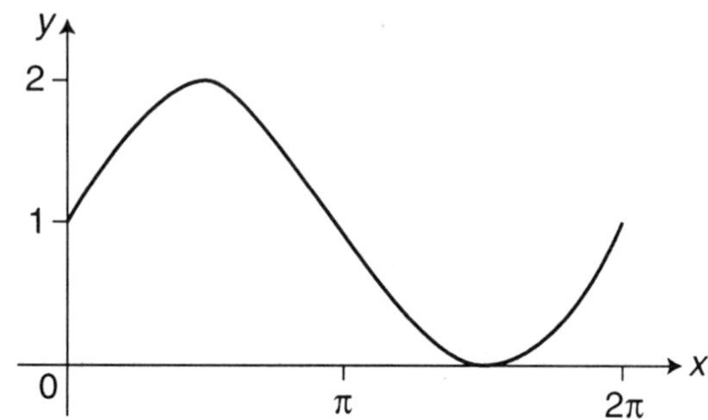

6. Write down the **exact** value of (i) $\cos 60°$ (ii) $\tan \frac{3\pi}{4}$.

7. State the condition for the curve $y = f(x)$ to have a stationary point at $(a, f(a))$.

8. Integrate (i) $2\sin x \cos x$ (ii) $2\cos^2 x - 1$.

SET 120

1. Simplify (i) $2x^0$ (ii) $(2x)^0$.

2. A set of lines passes through (0, 4). Write down an equation which is valid for all of these lines.

3. The roots of a quadratic equation are real. Describe the discriminant.

4. Copy and complete (i) $\log_a \frac{1}{n} = \ldots$ (ii) $\log_a X^n = \ldots$

5. Write down the coordinates of the centre and length of the radius of the circle $x^2 + y^2 + 2hx + 2ky + l = 0$.

6. Two vectors **u** and **v** are such that **u . v** = 0. Give three possible interpretations of this result.

7. Is $\frac{1}{3}\mathbf{i} - \frac{2}{3}\mathbf{j} + \frac{2}{3}\mathbf{k}$ a unit vector? Explain.

8. Integrate (i) $(ax + b)^n$ (ii) $\sin(ax + b)$

ANSWERS

SET 1

1. 9
2. $x^3 + 3x^2 - 6x - 8$
3. 5
4. (1, 3)
5. 11
6. 0, 6
7. 7·66 cm^2
8. (0, 1) (2π, 1) etc.

SET 2.

1. (i) ±2 (ii) –3
2. $3y(3x + 2y)(2x - 5y)$
3. $x + 2 + \frac{1}{x}$
4. $\frac{-5}{7}$
5. 10 units^2
6. 7
7. $\tan x$
8. $\frac{-\sqrt{6}}{3}$

SET 3

1. $x + 1$
2. $(x - 1)(x + 1)(x - 2)(x + 2)$
3. (i) $\sqrt{3}$ (ii) 4
4. (i) $y = 6$ (ii) $x = 5$
5. 130°
6. {–2, –1, 0, 1, 2}
7. 78·5°
8. 130

SET 4

1. (i) 5 (ii) 1
2. (i) $2x(x - 4y)$ (ii) $(x - 6)(x + 1)$ (iii) $x(x + 1)(x - 2)$
3. $1 + x$
4. 36 units^2
5. $3 < c < 13$
6. $x > \frac{4}{5}$
7. 4·5 cm
8. $\left(2\frac{1}{2}, 5\right)$

SET 5

1. $\frac{3}{4}y^{1/6}$
2. $x^3 + 4x^2 + x - 6$
3. $2\frac{1}{2}$
4. $4\sqrt{2}$
5. (i) $y = 3x + c$ (ii) $y = mx + 5$
6. $x = 2$, $y = -3$
7. $\left\{\theta : \frac{\pi}{6} < \theta < \frac{5\pi}{6}\right\}$
8. sin 42°; cos 48°

SET 6

1. $4\frac{1}{2}$
2. $2(4x^2 + 13xy + 8y^2)$
3. $2\sqrt{3}$
4. (1, –1)
5. $\frac{-7}{4}$
6. {–4, 9}
7. Max = 6; min = 4
8. $\frac{3}{4}$ units^2

SET 7

1. $(6b + 7l - 42)$ cm^2
2. $a^2 + ab - ac - 2b^2 + bc$
3. 1; –1
4. Collinear
5. $x = -4\frac{1}{2}$, $y = -4$
6. 4 units^2
7. $\frac{x(x - 2)}{3(3x - 1)}$
8. 5 mm^2

SET 8

1. –5
2. $\sqrt{120}$
3. $6x^3 + 7x^2 - 23x - 30$
4. 2
5. $\frac{-\sqrt{3}}{2}$
6. 0, 2
7. $-\sin^2 A$
8. (i) $y = 4$ (ii) $x = 3$ (iii) $4x - 3y = 0$

SET 9

1. $x^2 - 11x + 5$
2. $\dfrac{1}{\sqrt{2}}$
3. $\dfrac{x-3}{2}$
4. 27
5. $\dfrac{3v - 2\pi r^3}{3\pi r^2}$
6. (i) 6·3 cm (ii) 18·8 cm^2 (iii) 15·6 cm^2
7. 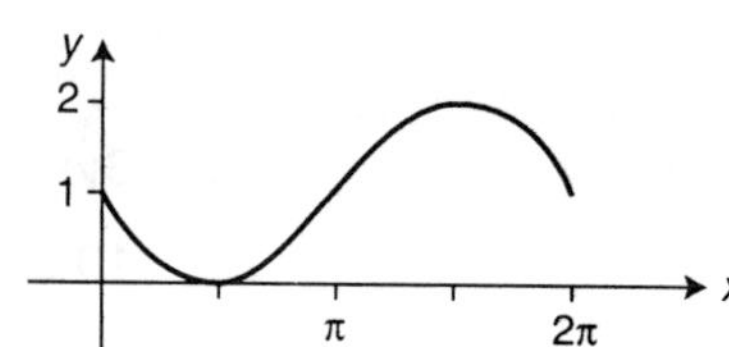

8. (–2, 0)(3, 0)

SET 10

1. $-12x$
2. ± 3
3. 155
4. $(a - c)(a - 2b + c)$
5. $(p, 0)$
6. (a) $r = \dfrac{v-u}{t}$ (b) $F = \dfrac{9C}{5} + 32$ (c) $g = \dfrac{u^2 - v^2}{2s}$
7. (4, 9)
8. 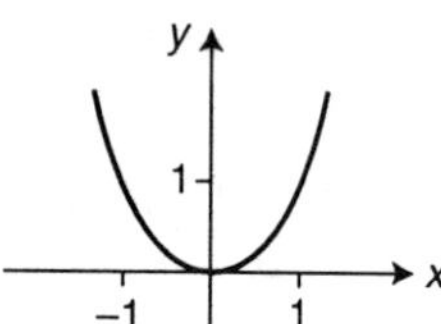

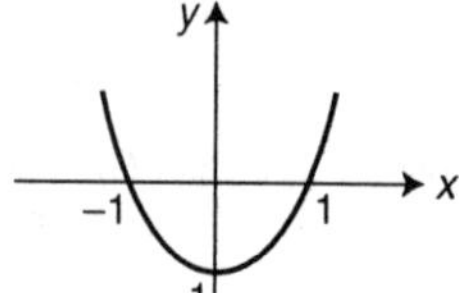

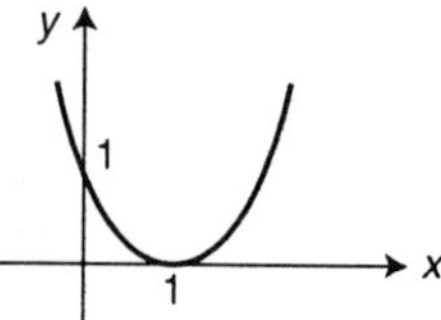

SET 11

1. 1
2. $\dfrac{4 - 2\sqrt{3}}{\sqrt{3}} = \dfrac{4\sqrt{3} - 6}{3}$
3. 29°
4. 0, 5
5. 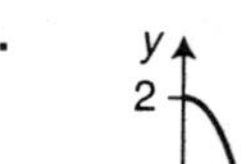
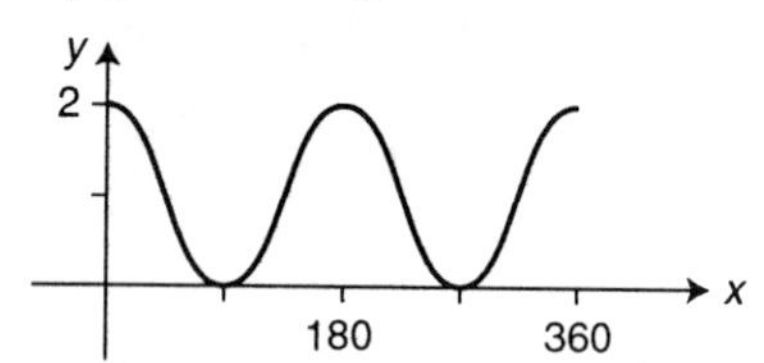

6. –2, 16
7. –2
8. $a^{n^2 - 2n}$

SET 12

1. –2·32, 4·32
2. (i) ±2 (ii) 2, 3
3. 60
4. $\dfrac{u+t}{2u}$
5. 30, 210
6. (i) (3, 0) (ii) $\left(1\frac{1}{2},\ 2\frac{1}{4}\right)$
7. $x = 1\frac{1}{2}$
8. (i)
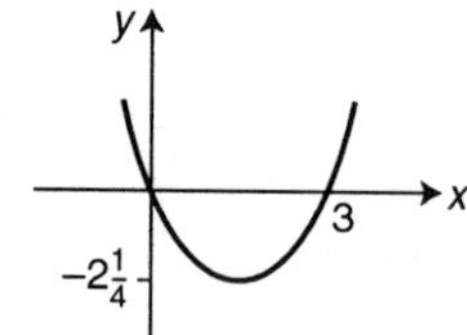

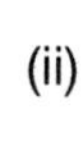(ii)
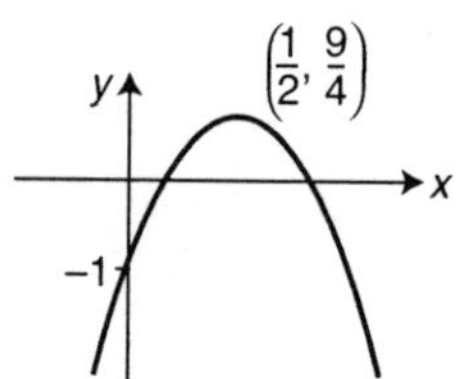

SET 13

1. $-\dfrac{1}{2}$
2. 76°
3. $4x - 3y - 2 = 0$
4. $\dfrac{-3}{2}$
5. $3x - 2y = 13$
6. –8
7. $m_{AB} = -1 = m_{AC}$
8. $x - 2y + 9 = 0$

SET 14

1. $p = 2$
2. 10 units^2
3. $5x - 3y - 9 = 0$
4. $m_{AB} = \frac{2}{3} = m_{AC}$
5. $3x + 2y = 4$
6. (8, 6) is on all lines
7. $x - 2y = 6$
8. $2\frac{2}{3}$

SET 15

1. $x + y = 2$
2. $-\frac{1}{2}$
3. $3x - y + 5 = 0$
4. (3, –2)
5. $-\frac{1}{5}$
6. $x + y = 10$
7. $-\frac{11}{2}$
8. $m_{AB} \cdot m_{CD} = \left(\frac{3}{4}\right)\left(-\frac{4}{3}\right) = -1 \Rightarrow AB \perp CD$

SET 16

1. 53·1°
2. $m = -\frac{5}{4}$; 5
3. $x + 2y = 11$
4. (–1, 3)
5. $2x + 3y = 9$
6. –7
7. $m_{AB} \cdot m_{AC} = -1$
8. $x - y - 5 = 0$

SET 17

1. $h = 5$
2. $x - y + 2 = 0$
3. $m_{AB} = m_{BC} = 2 \Rightarrow$ collinear
4. –1
5. $\frac{1}{2}, -3$
6. $3x - 2y - 7 = 0$
7. $m_{PQ} \cdot m_{RS} = (-2)\left(\frac{1}{2}\right) = -1 \Rightarrow PQ \perp RS$
8. $x - 7y - 19 = 0$

SET 18

1. (i) $3x + y = 0$ (ii) $y = 3$ (iii) $x = -1$
2. $m_{PQ} = \frac{1}{3} = m_{RS} \Rightarrow PQ \| RS$
3. $3x - 2y = 0$
4. (1, –2)
5. $PQ = \sqrt{13} = QR \Rightarrow$ isosceles
6. $\frac{1}{3}$
7. $m_1 \cdot m_2 = \frac{-2}{3} \cdot \frac{3}{2} = -1 \Rightarrow$ perpendicular
8. $3x - 2y + 6 = 0$

SET 19

1. 0
2. 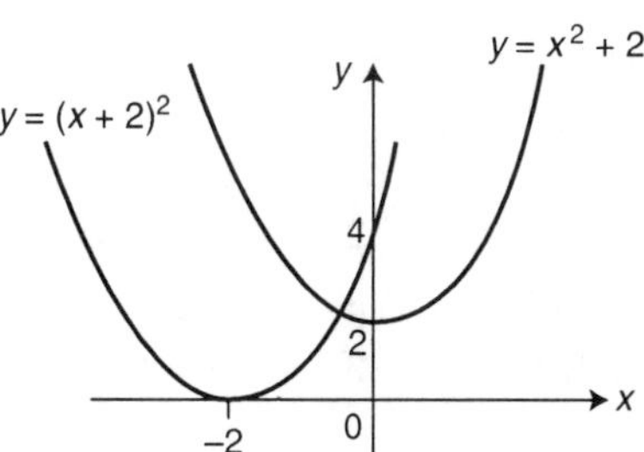

3. (i) π (ii) π (iii) 4π
4. $\frac{1}{4}$
5. 6
6. –2
7. sin $2x$
8. $x^2 + 8x + 15$

SET 20

1. $10x^2 - 3x$
2. 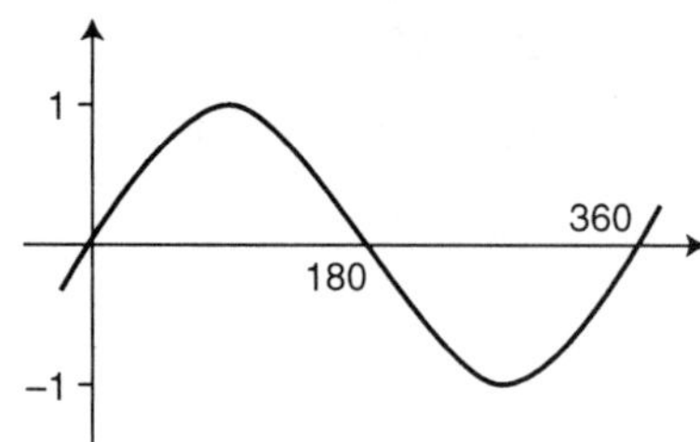

3. $x^2 + 6x + 8$; $x^2 + 2$
4. $8h$
5. ± 2
6. 16
7. 300
8. $\sin^2 x$

SET 21

1. 0
2. $(\pm 2, 2)$
3. $1 + \cos x$
4. (i) 360 (ii) 180
5. $4x^2 + 4x + 1$
6. 5, −3
7. $x^2 + x$
8. 0

SET 22

1. (i) $\{x : -3 \leq x \leq 3\}$ (ii) $\{y : 0 \leq y \leq 3\}$
2. $6x + 3$
3. (i) 90° (ii) 1440°
4. $(x - 2)^2 - 3$
5. $y = (x - 2)^2$
6. $x\sin x$
7. $x^2 + 6x + 12$
8. 12; 305

SET 23

1. (i) $\dfrac{\sqrt{3}}{2}$ (ii) $\dfrac{\sqrt{3}}{2}$ (iii) 1
2. $x^2 - x + 1$
3. 3, −1
4. 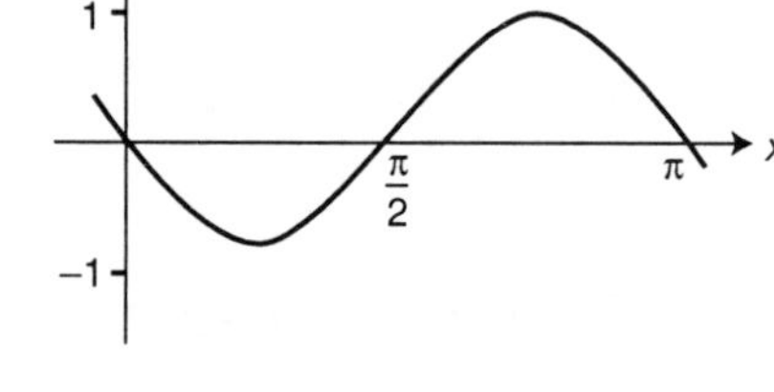

5. 1
6. $-4a$
7. (i) $\{x : x \geq 0\}$
   (ii) R
   (iii) $\{x : x \geq 3\}$
   (iv) $\{x : x \geq 0\}$
8. 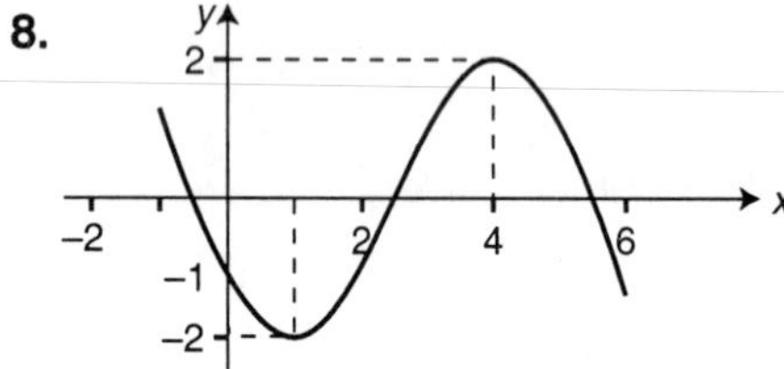

SET 24

1. $\dfrac{1}{2}$
2. $4x - 3$; $4x^2 - 4x$
3. 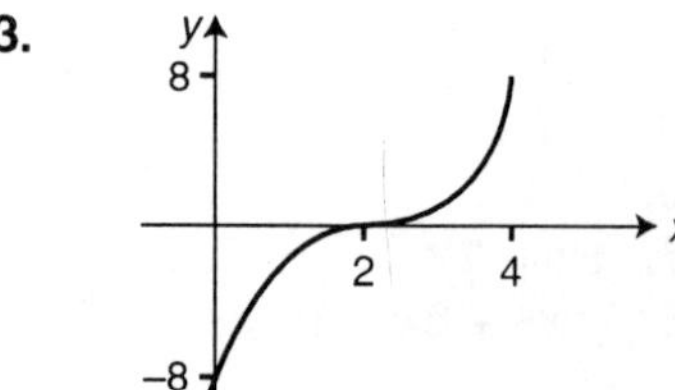

4. 7
5. −3, 2
6. $y = 2x^2 - 8x + 6$
7. $2(x - 3)^2 - 1$
8. (i) 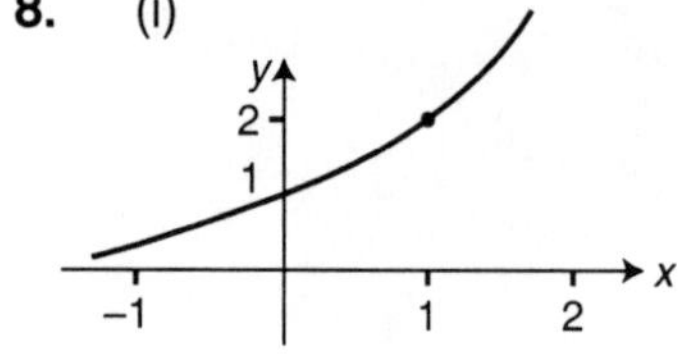

(ii) 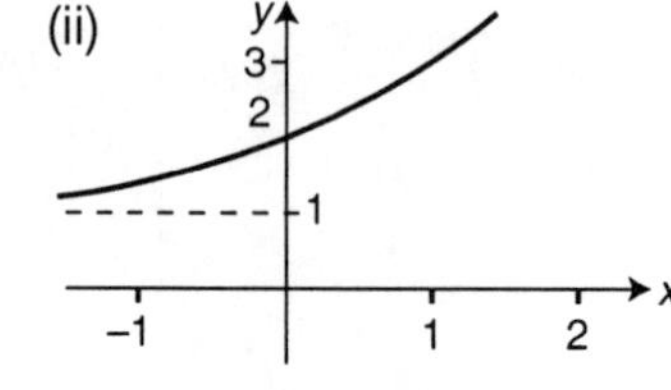

(iii) 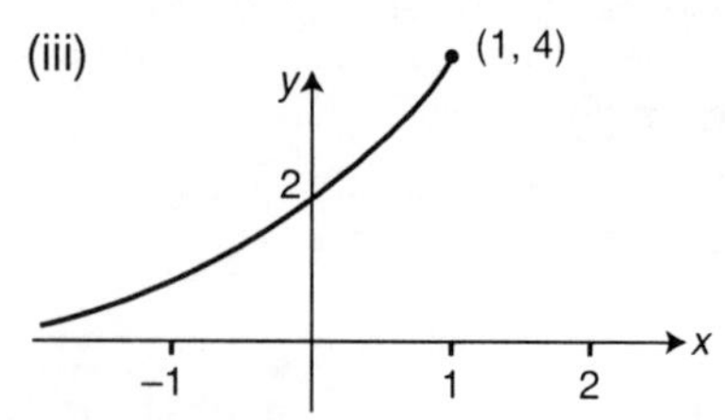

SET 25

1. -12 **2.** $12x - 45x^2$ or $3x(4 - 15x)$ **3.** -6

4. 6 **5.** $V'(r) = \pi r^2$ **6.** 45°

7. $\frac{1}{2x^{1/2}} - \frac{1}{2x^{3/2}}$ **8.** 1

SET 26

1. 2 **2.** $\frac{3}{2}x^{1/2}$ **3.** 0

4. $\frac{dy}{dx} = 3x^2 \geq 0$ **5.** $4 - \frac{3}{x^2}$ **6.**

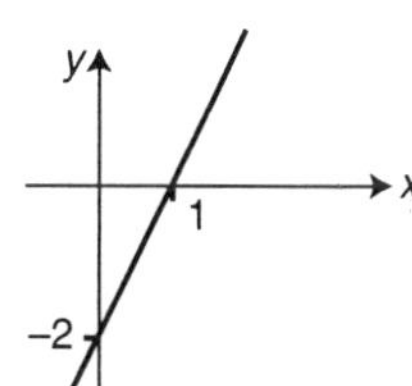

7. $x = 1$ **8.** $(1, -1)$ min t.pt.

SET 27

1. -10 **2.** (i) $\frac{1}{2\sqrt{u}}$ (ii) $\frac{-2}{u^3}$ **3.** 4

4. $x > 2$ **5.** 2 **6.** $\frac{3}{2}x^{1/2} + x^{-1/2} + \frac{3}{2}x^{-3/2}$

7. $f(a)$ and $f(b)$ **8.** $a = -1$, $b = 5$

SET 28

1. $\frac{1}{4}$ **2.** $3kx^2 - 2kx$ or $k(3x^2 - 2x)$ **3.** 5

4. 39 **5.** 2 **6.** $1 + \frac{1}{x^2}$

7. $x < -1$ and $x > 1$ **8.** $(1, 5)$

SET 29

1. -12 **2.** $\frac{3}{2}p^{1/2}$; $\frac{1}{2}p^{-1/2}$ **3.** -9

4. $f'(x) = -6x^2 \leq 0$ **5.** $4 - 12x$ **6.** -1

7. $3 - 6x^{-1/2}$ **8.** $x < 1$

SET 30

1. $2\pi r + 2\pi h$ **2.** $13\frac{1}{2}$ **3.** $f(0)$, $f(a)$

4. $4x^3 + 12x$ **5.** -5 **6.** $\frac{3}{2}x^{1/2} - \frac{3}{2}x^{-1/2}$

7. $(1, 0)$ **8.** Falling point of inflection at $(0, 4)$

SET 31

1. 4; 31·5 **2.** (*a*); 0·6 < 1 **3.** 4

4. 8 **5.** 2 **6.** u_5

7. 12 **8.** $50(0{\cdot}8)^6 = 13{\cdot}1$ g

SET 32

1. 15; 0 **2.** 6 **3.** 20

4. 0·2 **5.** 3 **6.** u_5

7. $p = 3q$ **8.** $u_{n+1} = 0{\cdot}3u_n + 600$

SET 33

1. $\left(3, 1\frac{1}{2}\right)$ 2. 20 3. $3x - y = 7$
4. $4\pi r + 2\pi h$ 5. (i) $3(2x + 1)^2$ (ii) $6x^2 + 1$
6. AB = BC; $m_{AB} \,.\, m_{AC} = -1$ or $AC^2 = AB^2 + BC^2$
7. 4 8. $1 - \frac{4}{x^2}$

SET 34

1. $\frac{-1}{2u^{3/2}}$ 2. -5 3. 6·25
4. $f'(a)$ 5. $x > 1$ 6. -6, 10
7. $2 - (x - 3)^2$ 8. (i) $y = x$ (ii) $y = 2x$

SET 35

1. 0 2. 2 3. $\frac{2}{5}$
4. (i) -2 (ii) -2 (iii) -2 5. 2, 11, 101, 1001, 10 001 6. (0, 0) pt. of rising inflection
7. 1 8. (i) -4 (ii) $-\frac{2}{3}$

SET 36

1. 0 2. $\left(7\frac{1}{2}, -1\frac{1}{4}\right)$
3. (i) $y = 3x + 2$ (ii) $4x + 5y = 10$ 4. $f'(a)$
5. 20 6. $\frac{1}{2}u^{-1/2} - \frac{9}{2}u^{1/2}$ 7. $-1 < x < 1$
8. (i) 91 (ii) 9

SET 37

1. -1, 4
2. $4x(1 - x)$
3. 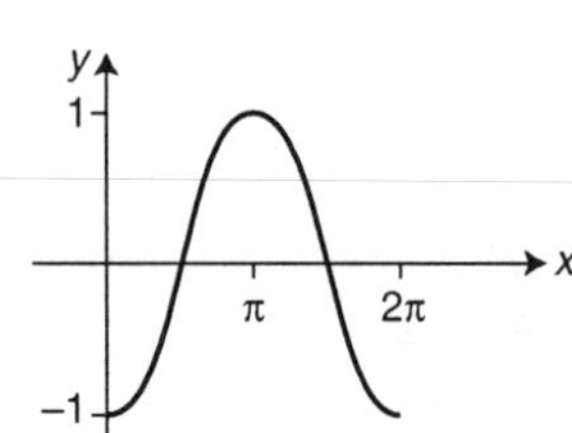

4. (2, 2)
5. $3x + 4y = 0$
6. $\frac{a - 1}{a - 4}$
7. (i) $2(x - 1)$ (ii) $2x$
8. $(-1, 3)$

SET 38

1. $x = -1$ 2. $-1, \frac{1}{3}$ 3. 12·5
4. $\frac{-4(x + 1)}{x^2(x + 2)^2}$ 5. 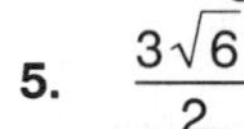 $\frac{3\sqrt{6}}{2}$
6.  (i)
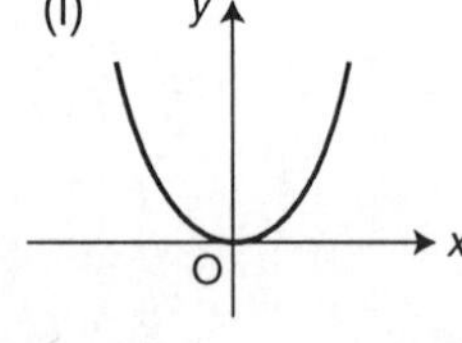

(ii)
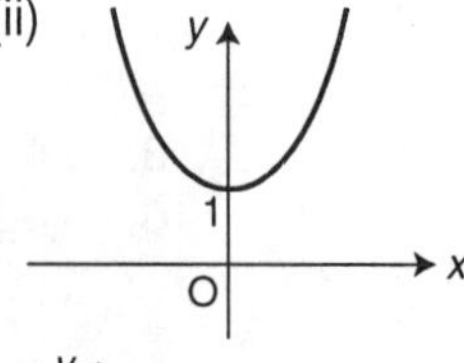

(iii)
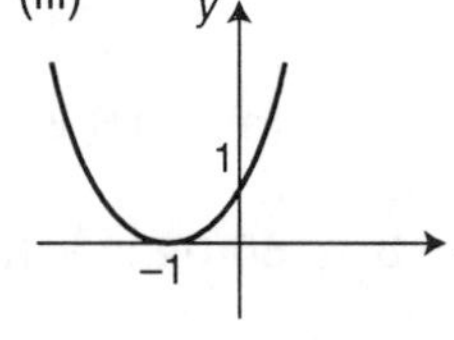

(iv)
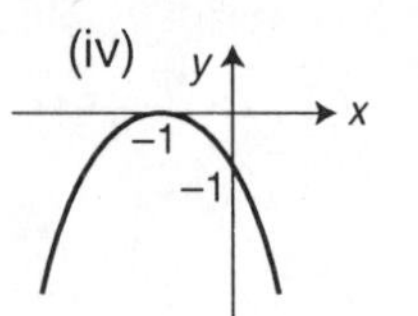

(v)
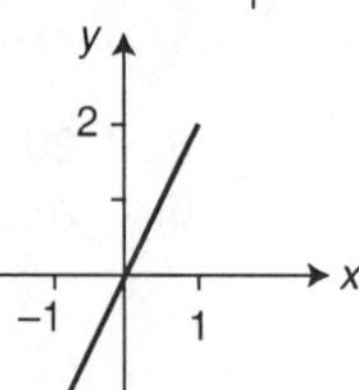

7. $4x^2 + 6x + 2$
8. $3x + 2y - 17 = 0$

SET 39

1. $f = \dfrac{uv}{v+u}$
2. $\dfrac{3}{2}z^{1/2} - \dfrac{1}{2}z^{-3/2}$
3. 8, 4
4. (i) 68 (ii) 6
5. $\dfrac{-5}{3}$
6. $x - 7y + 8 = 0$
7. (0, 5), (–2, 13)
8. (i)

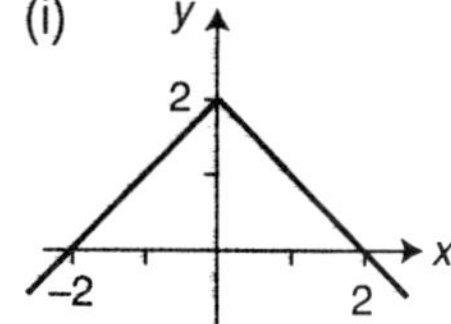

(ii) not one to one, e.g., $f(-2) = 0 = f(2)$

SET 40

1. $x^3 - \dfrac{1}{x^3}$
2. $2(x + 4)$ or $2x + 8$
3. $\dfrac{13}{15}$
4. $(3 - 2x)(7 - 6x)$
5. (i) –4 (ii) 27
6. 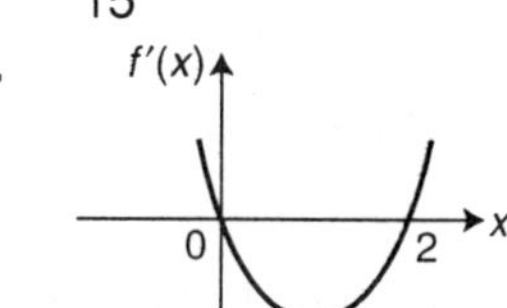

7. $9x - y + 6 = 0$
8. 4

SET 41

1. $x = 2$, $y = -1$
2. 13
3. $a = 1$, $b = 1$, $c = 2$
4. x^4
5. –1, 5
6. $u_5 = 251$
7. 4 m s^{-1}; $2\frac{1}{2}$ s
8. $L\hat{M}K = 90°$

SET 42

1. $5x(3x - 7)$
2. $\dfrac{9}{2}u^{1/2} + u^{-1/2}$
3. 0, 180, 360
4. 1·5
5. 4
6. $y = 4 + 2x - 2x^2$
7. (–3, 9), $\left(\dfrac{1}{3}, \dfrac{13}{27}\right)$
8. $6x - y = 11$

SET 43

1. $x = \dfrac{1}{5}$
2. 4
3. 1, –5
4. (i) –3 (ii) 8
5. $\dfrac{1}{x^2 - 1}$
6. (i)

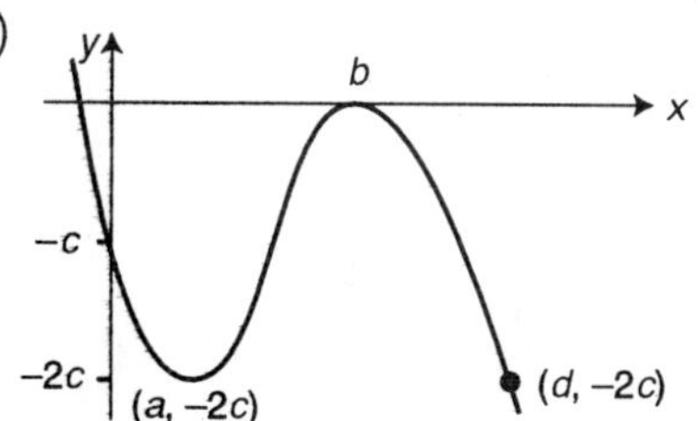

(ii)

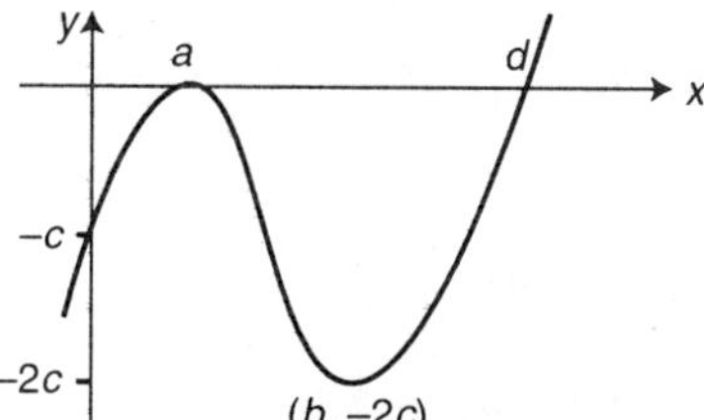

7. (–1, 2) max, (1, –2) min
8. $2x + 3y = 10$

SET 44

1. (i) $3x^2(x - 2)(x + 2)$ (ii) $(5\cos x° + 1)(\cos x° - 1)$
2. $4 - \dfrac{1}{x^2}$
3. $\dfrac{2\sqrt{6} - \sqrt{2}}{4}$
4. $13\frac{1}{2}$ units^2
5. $\left(x + 2\frac{1}{2}\right)^2 - 7\frac{1}{4}$
6. $10x - y = 14$
7. (–1, 2)
8. Settles at 150

SET 45

1. 4; –2 **2.** 16 **3.** 27, $4x^2 + 6x + 16$
4. 18 **5.** –3 **6.** $(x-2)(x+2)(x-3)$
7. $x(x-2)(x+1)$ **8.** –1

SET 46

1. $\frac{7}{8}$ **2.** $(x+1)(x-2)(x+5)$ **3.** $(x-1)(x-2)(x+3)$
4. $f(0) = 1,\ f(1) = -4 \Rightarrow f(x) = 0$ for some x between 0 and 1
5. 1, –1, –2 **6.** 17 **7.** 3, 4, $2\frac{1}{2}$ **8.** $\frac{2}{3}$, 1, –3

SET 47

1. $a = 6,\ p = 3$ **2.** $k = -1\frac{1}{2},\ m = 2\frac{1}{4}$ **3.** $x = -1$
4. $\{x : x < 1\} \cup \{x : x > 2\}$ **5.** $\left\{x : x \le \frac{-2}{3}\right\} \cup \left\{x : x \ge 2\frac{1}{2}\right\}$
6. $-4 < x < 2$ **7.** $(x+3)^2 - 1$ **8.** $8 - (x+1)^2$

SET 48

1. Real (rational), distinct **2.** Real (rational), distinct **3.** 16
4. $b^2 = a^2 \Rightarrow b = \pm a$ **5.** 0, 5 **6.** $k \le \frac{4}{3}$
7. $\Delta = k^2 + 20 \ge 20 > 0 \Rightarrow$ real roots **8.** $\Delta = 0$

SET 49

1. 305 **2.** –2 **3.** –6
4. Unreal **5.** $f(-2) < 0,\ f(-1) > 0$ **6.** ±1
7. $\frac{-1}{2}$, 1 **8.** $p^2 - 28 \ge 0 \Rightarrow 4p^2 - 96 \ge 16 > 0$

SET 50

1. –27 **2.** $x = 2$ **3.** $a = 1$
4. $2(x-3)(x+4)(2x+1)$ **5.** one **6.** $x^2 + 2x$; 10
7. 100 **8.** $a^2 + (3-k)a + (4-2k) = 0$

SET 51

1. $f'(x) = 2x;\ \frac{1}{3}x^3 + c$ **2.** $x^4 - x^2 + x + c$ **3.** $y = x$
4. $s = t^3 - t - 3$ **5.** $y = x^2 + x + 1$ **6.** $4x - x^2 + c$ **7.** $y = 2x - 4$
8. (i) $18x - 6$ (ii) $3x^3 - 3x^2 + x + c$

SET 52

1. (i) $-2x$ (ii) $c + 9x - \frac{1}{3}x^3$
2. $\frac{1}{3}x^3 + 2x^2 + 4x + c$
3. $\frac{1}{2}x^4 - \frac{1}{2}x^2 + c$
4. $\frac{4}{3}x^3 - 2x^2 + x + c$
5. $c - \frac{1}{x}$
6. 0

7.

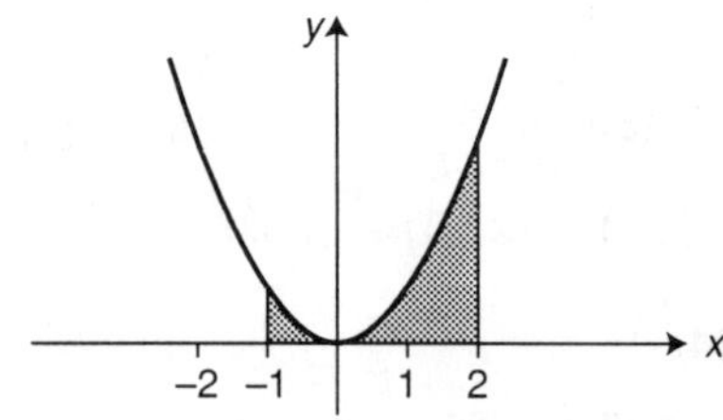

8. $-k$

SET 53

1. $7\frac{1}{3}$ units^2
2. $2x^3 - x^2 + c$
3. $\frac{-3}{2}$
4. 21
5. $\frac{4}{3}x^3 + 2x^2 + x + c$
6. $\int_{-4}^{2} (f_2(x) - f_1(x))\, dx$
7. $s = t^3 - t^2 + 4$
8. $y = 2x^3 - x^2 + 5$

SET 54

1. $\frac{2}{5}$
2. 0
3. 4
4. 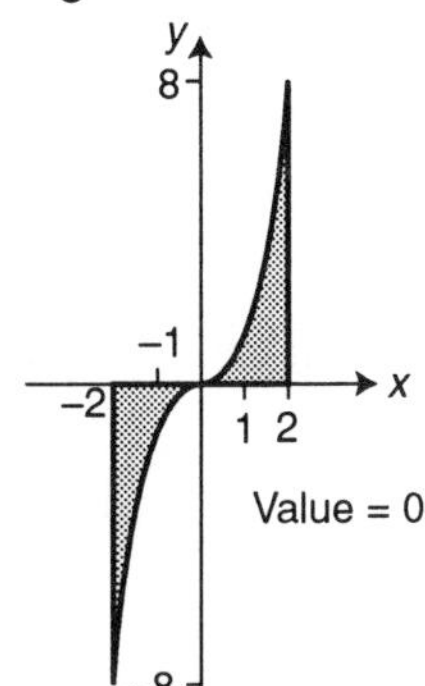

5. 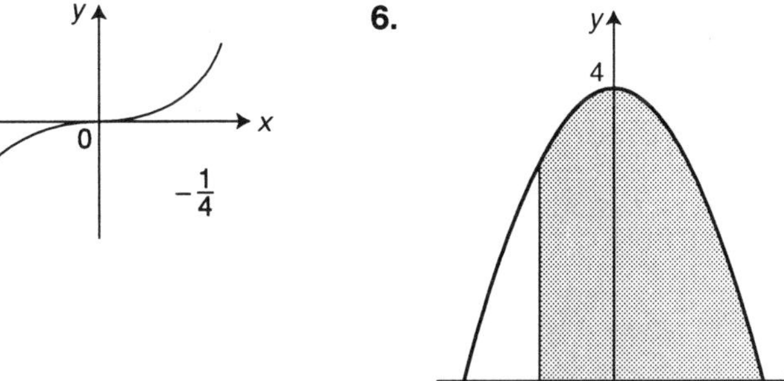

7. $18\frac{3}{5}$
8. $18\frac{2}{15}$

SET 55

1. $k + z$
2. (i) $x^2 + x + c$ (ii) $\frac{2}{3}x^3 + \frac{1}{2}x^2 + c$
3. 2
4. $\frac{5}{3}$
5. $17\frac{1}{3}$
6. $f(x) = x^2 - 3x + 4$
7. 39
8. $\int_{0}^{a} (g(x) - h(x))\, dx$

SET 56

1. $x^2 + x; \frac{1}{3}x^3 + \frac{1}{2}x^2 + c$
2. $\frac{1}{2}x^4 - x^3 + \frac{1}{2}x^2 - x + c$
3. $x + \frac{1}{x} + c$
4. (2, 0), (3, 0)
5. $\int_{2}^{3} (6 - 5x - x^2)\, dx$
6. $y = x^2 + 3x + c$
7. $y = x^2 + 3x + 1$
8. 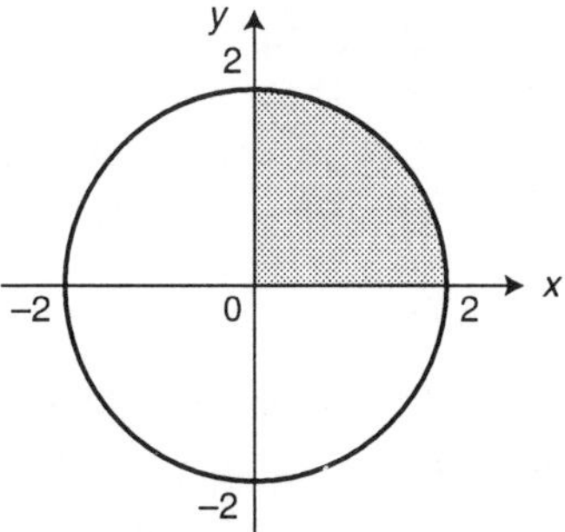

SET 57

1. {131·8, 228·2}
2. {166, 226}
3. $0, \frac{\pi}{3}, \pi$
4. (i) sin x (ii) 1
5. (i) 0 (ii) 1
6. 45°
7. 13 cm; 13·3°
8. 41·4°

SET 58

1. 10
2. $\frac{12}{13}$
3. $\sin 3y$
4. (i) 6 cm (ii) 76·4°
5. (i) cos 65° (ii) $\sin 135° = \frac{1}{\sqrt{2}}$ (iii) $-\cos 70° = \cos 110°$
6. $\frac{\sqrt{3}-1}{2\sqrt{2}} = \frac{\sqrt{6}-\sqrt{2}}{4}$
7. {15, 165, 195, 345}
8. $\frac{\pi}{4}$

SET 59

1. (i) $\cos 2x$ (ii) 1
2. $(2\cos x - 1)(\cos x + 1)$
3. $\sin x$
4. $\cos A$
5. $\sin (A - C)$
6. 21·8°
7. 50·5°
8. 63·4°

SET 60

1. 29
2. 80, 200
3. $2\sin \theta$
4. 30, 150, 210, 330
5. 30, 90
6. (i) 70·5, 90, 270, 289·5 (ii) 90
7. 180, 360
8. 71·6°

SET 61

1. $2\sin A \cos B$
2. (i) 0 (ii) 0
3. $\cos 2A$
4. 25
5. $\cos 2A$
6. 90°
7. 54·5°
8. 35·3°

SET 62

1. $\frac{24}{25}$
2. 55·6°
3. 1
4. $\frac{-3}{5}$
5. 45°
6. $22\frac{1}{2}, 67\frac{1}{2}, 112\frac{1}{2}, 157\frac{1}{2}$
7. $\frac{-16}{65}$
8. 0

SET 63

1. ±3
2. {(–1, –1), (1, 1)}
3. Outside
4. (4, 2); $\sqrt{29}$
5. 2
6. $x^2 + y^2 - 6x - 6y + 5 = 0$
7. $(x-1)^2 + (y+2)^2 = 9$
8. $\frac{4}{3}$

SET 64

1. (4, –1)
2. (2, –3), 4
3. $x^2 + y^2 - 8y + 7 = 0$
4. $x^2 + y^2 \pm 4x \pm 4y = 0$
5. (2, –1), 4
6. $x^2 + y^2 - 6x - 2y - 6 = 0$
7. $4x - 3y + 25 = 0$
8. $x^2 + y^2 - 14x - 4y + 37 = 0$

SET 65

1. $r^2 < 0$
2. 7
3. (2, –3), 4
4. $\left(1, \frac{-7}{4}\right)$
5. $x^2 + y^2 + 2x - 4y - 4 = 0$
6. (1, 2)
7. 2 units left, 1 unit up
8. $(x-5)^2 + (y-3)^2 = 25$

SET 66

1. $\frac{-2}{3}$
2. $(x-5)^2 + (y-1)^2 = 13$
3. $(x-1)^2 + (y-2)^2 = 18$
4. (6, 6), (–2, –2)
5. (1, 1); $\sqrt{5}$
6. $2x + y - 8 = 0$
7. $k < 13$
8. (i) $AB = r + s$ (ii) $AB = r - s$

SET 67

1. $\left(\frac{3}{2}, \frac{-9}{4}\right)$
2. 0
3. $x^2 + y^2 - 4x + 2y - 4 = 0$
4. $(x-4)^2 + (y-2)^2 = 29$
5. Inside
6. $(-1, 2)$; $2\sqrt{5}$
7. $\frac{1}{2}$
8. $2x + y - 10 = 0$

SET 68

1. $x^2 + y^2 + 2gx + 2fy = 0$
2. $(1, 2)$; 5
3. $(9, 8)$
4. $(x-9)^2 + (y-8)^2 = 25$
5. $(13, 0)$
6. 7
7. $(x \pm 4)^2 + (y-5)^2 = 25$
8. Yes at $(-1, 3)$

SET 69

1. $\frac{2}{\sqrt{5}}\left(= \frac{2\sqrt{5}}{5}\right)$
2. $2(x+1)(x-2)(x+3)$
3. $4x + 3y = 25$
4. 21·8°
5. $6\frac{1}{3}$ units^2
6. $y = 5x^3 + 6x^2$
7. $(x-3)^2 + (y-4)^2 = 25$
8. $(x^2+1)^2 + 4$

SET 70

1. (i) $2\cos^2\frac{A}{2} - 1$ (ii) $1 - 2\sin^2\frac{A}{2}$
2. $a\sin\theta$
3. $x^2 + 2x$; 10
4. 70·5°
5. Real and distinct
6. $(x \pm 5)^2 + (y-4)^2 = 25$
7. $13\frac{1}{3}$
8. $\frac{1}{2}x^2 + 3x + c$

SET 71

1. $0, \frac{\pi}{3}, \pi, \frac{5\pi}{3}, 2\pi$
2. Distinct, real (rational)
3. (graph)
4. (Proof)
5. $(x+2)(x^2+4x+5)$
6. Yes
7. $\frac{2}{3}x^{3/2} + 2x^{1/2} + c$
8. No

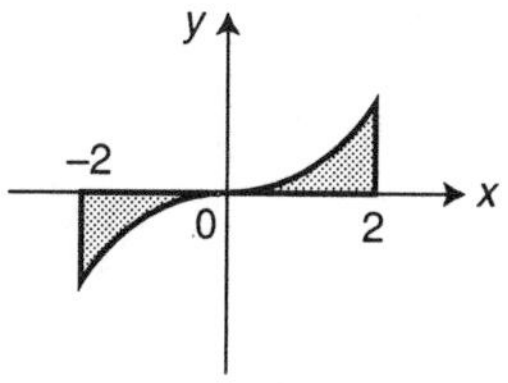

SET 72

1. 24 cm
2. $(x-2)(x+5)(2x-1)$
3. 45°
4. $\frac{\sqrt{3}+1}{2\sqrt{2}} = \frac{\sqrt{6}+\sqrt{2}}{4}$
5. $2x^2 + (2p-3)x + (p^2+3q) = 0$
6. $(x-7)^2 + (y-12)^2 = 29$
7. $y = x^4 - 3x^2 - 5$
8. $2\frac{2}{3}$ units^2

SET 73

1. 5; -3
2. $\{k < -2\} \cup \{k > 2\}$
3. $\frac{1}{2}$
4. $4\frac{2}{3}$
5. (i) $(2\sin x - 1)(\sin x + 3)$ (ii) $\sin A(2 - \sin A)$
6. $r = 4$ s
7. (graph)
8. $x = 3$

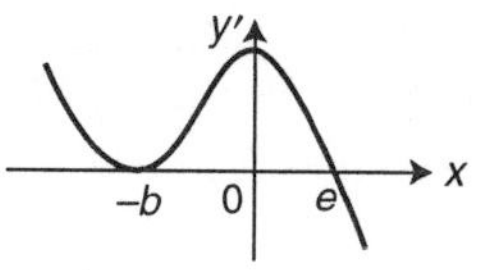

SET 74

1. $y = (x-2)^2$
2. $\{x : -3 < x < 8\}$
3. (i) $3y = 2x$ (ii) $(3, 2)$
4. $x^2 + y^2 = 13$
5. 1
6. 1
7. $x + y = 8$
8. 12 days

SET 75

1. 45°
2. $\cos 2x$
3. $(\cos A - 1)(\cos A - 2)$
4. 60°
5. $u_1 = 17,\ u_2 = 47,\ u_3 = 137,\ u_4 = 407$
6. $\frac{3}{2}x^{1/2} + x^{-1/2} - \frac{1}{2}x^{-3/2}$
7. $2\frac{5}{6}$
8. $(-2, -4), (4, 2)$

SET 76

1. 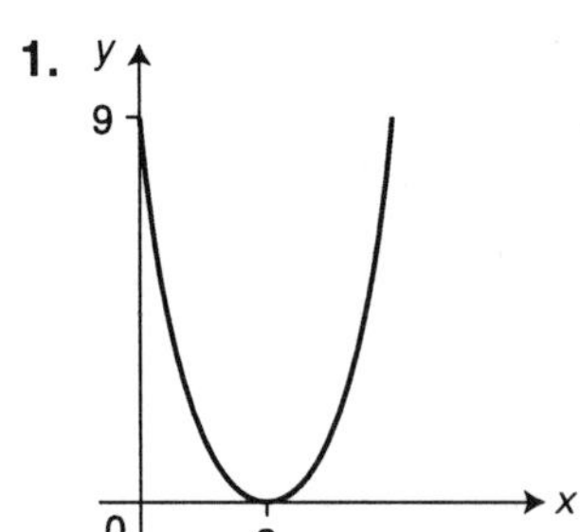

2. (i) $\frac{\sqrt{2}}{\sqrt{3}} = \frac{\sqrt{6}}{3}$ (ii) $\frac{1}{3}$
3. $x^{2/3} - 2 + x^{-2/3}$; $\frac{2}{3}x^{-1/3} - \frac{2}{3}x^{-5/3}$
4. 30°
5. $u_{n+1} = 0{\cdot}25u_n + 100$,
   u_n = number of mg of pollen on the nth morning
6. $\frac{1}{5}x^5 + \frac{2}{3}x^3 + x + c$
7. $5y = 2x + 29$
8. *(a)*

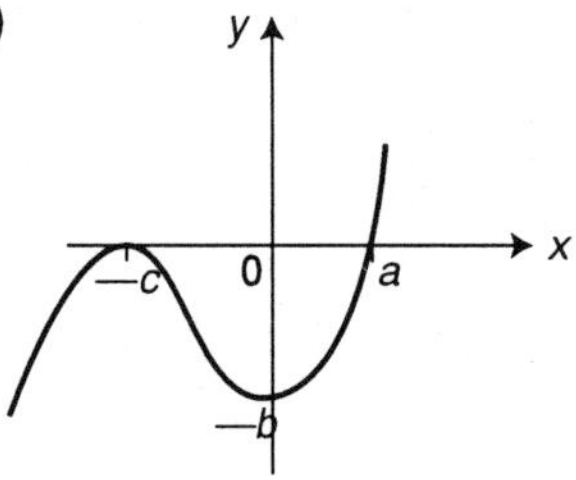

(b)

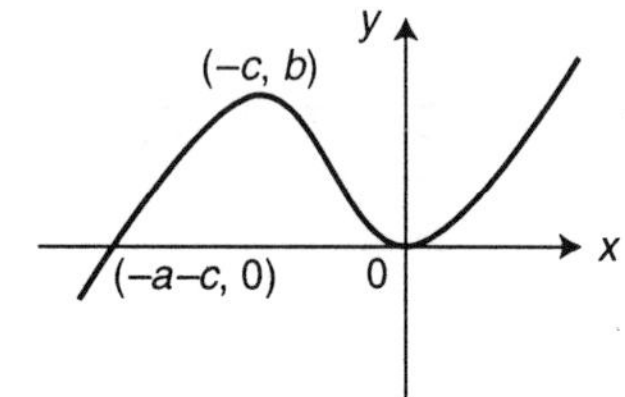

SET 77

1. $x = -1\frac{1}{2}$

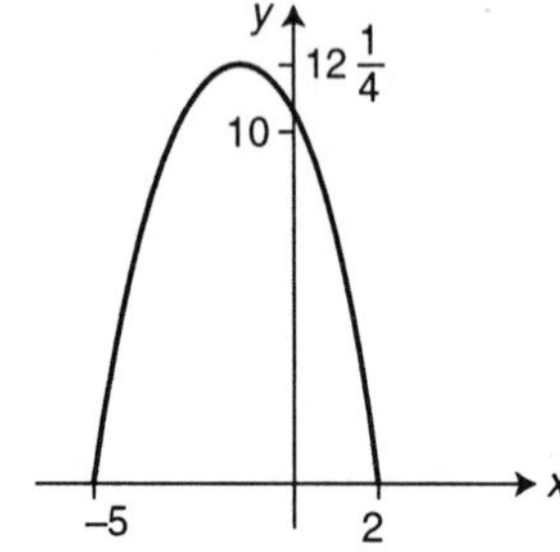

2. (i) 1 (ii) 1
3. $y = x$
4. 20
5. $\frac{-1}{x^2}$
6. $-1 < k < 7$
7. $2x^2 + 3$
8. $x^2 - 3x + 4$

SET 78

1. $(7, 1)$
2. $\frac{4\pi}{3}$ cm
3. $\sin^2 A$
4. $\{x : x \le -2\} \cup \{x : x \ge 6\}$
5. -9
6. $\{y : y \in R, y \ge 5\}$

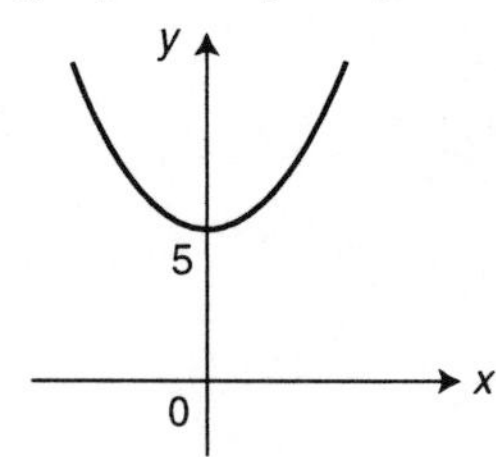

7. $4x + y = 11$
8. $2\frac{2}{3}$ units^2

SET 79

1. 1 unit right, 2 units up
2. 0
3. $2x^2 + (2c - 9)x + c^2 = 0$
4. $\frac{1}{2}$
5. 0, 2
6. $\frac{5}{6}$
7. $10\frac{2}{3}$ units^2
8. $(x - 2)^2 + (y - 6)^2 = 9$

SET 80

1. -1
2. 81
3. $\frac{1}{2}$
4. $x \geq -2$
5. $\frac{1}{x^2 - 1}$
6. 1·5
7. (3, 4)
8. $(x - 3)^2 + (y - 4)^2 = 25$

SET 81

1. (i) 25 (ii) 7
2. collinear
3. $\begin{pmatrix} 1 \\ -1 \end{pmatrix}$
4. (0, 5)
5. 16
6. 0
7. $\frac{4}{21}$
8. $\mathbf{p} + \mathbf{q} - \mathbf{r}$

SET 82

1. $\sqrt{61}$
2. $3a = 2b$
3. (i) (6, 9) (ii) Both = 13 or 2 diagonals of a parallelogram.
4. (2, 6, 1)
5. $-4\frac{1}{2}$
6. $a^2 - b^2$
7. $\frac{14}{15}$
8. $\frac{1}{4}(\mathbf{p} + \mathbf{q} - 2\mathbf{r})$

SET 83

1. $\sqrt{66}$
2. $\overrightarrow{XY} = 3\overrightarrow{YZ}$
3. (i) $\mathbf{b} - \mathbf{a}$ (ii) $\mathbf{m} = \frac{1}{2}(\mathbf{a} + \mathbf{b})$
4. (1, 4, 3)
5. (i) 1 (ii) 0
6. (i) 18 (ii) -7
7. 90°
8. $\begin{pmatrix} \frac{1}{2} \\ -4 \\ -1\frac{1}{2} \end{pmatrix}$

SET 84

1. 6
2. 6
3. $m = 6$, $n = -1$
4. (−1, 5, −6)
5. (i) $\frac{1}{2}$ (ii) $\frac{-1}{2}$
6. 12
7. $\frac{\overrightarrow{QP} \cdot \overrightarrow{QR}}{|\overrightarrow{QP}|\,|\overrightarrow{QR}|}$
8. $\frac{\sqrt{17}}{9}$

SET 85

1. $\begin{pmatrix} 2 \\ -3 \\ -4 \end{pmatrix}$

2. $\begin{pmatrix} 3 \\ 6 \\ -3 \end{pmatrix}, \begin{pmatrix} 1 \\ 2 \\ -1 \end{pmatrix}$

3. $\overrightarrow{AB} + \overrightarrow{AE} + \frac{1}{3}\overrightarrow{AD}$

4. $\begin{pmatrix} 17 \\ 3 \\ -1 \end{pmatrix}$

5. $\begin{pmatrix} 20 \\ -5 \\ -8 \end{pmatrix}$

6. $\begin{pmatrix} -3 \\ 8 \\ 7 \end{pmatrix}$

7. 6

8. Proof

SET 86

1. $\overrightarrow{AP} = \overrightarrow{AD} + \frac{1}{3}\overrightarrow{AB}$, $\overrightarrow{AQ} = \overrightarrow{AB} + \frac{3}{5}\overrightarrow{AD}$

2. $\begin{pmatrix} 8 \\ -6 \\ 10 \end{pmatrix}$

3. $\begin{pmatrix} 12 \\ 6 \\ 18 \end{pmatrix}$

4. 240

5. $\overrightarrow{VP} = \overrightarrow{AP} - \overrightarrow{AV}$

6. $\begin{pmatrix} -6 \\ -8 \\ 5 \end{pmatrix}$

7. $\begin{pmatrix} -2 \\ 4 \\ 13 \end{pmatrix}$

8. 73·0°

SET 87

1. $8x(x^2 + 4)^3$

2. $\dfrac{1}{(1 - x)^2}$

3. $\cos x - \sin x$

4. 3

5. 2π

6. $-2\cos x \sin x = -\sin 2x$

7. (i) $x^2 + x + c$ (ii) $\frac{1}{8}(2x + 1)^4 + c$ (iii) $\frac{1}{16}(2x + 1)^8 + c$

8. 1 unit^2

SET 88

1. $12x(2x^2 - 1)^2$

2. $\dfrac{-1}{2(x + 1)^{3/2}}$

3. $-2\sin(2x - 1)$

4. 0

5. 2 units^2

6. $4\cos x(1 + \sin x)^3$

7. $\frac{1}{2}(1 - \cos 2u)$; $\frac{1}{2}u - \frac{1}{4}\sin 2u + c$

8. $1\frac{1}{3}$ units^2

SET 89

1. $4(2x + 3)(x^2 + 3x - 5)^3$

2. $\dfrac{-4x}{(1 + x^2)^2}$

3. $2\cos 2x$

4. 3

5. $\pi + 2$ units^2

6. $3\sin^2 x \cos x$

7. $(3x + 4)^{1/3} + c$

8. $2\sqrt{2}$ units^2

SET 90

1. $\frac{9}{2}x^2(x^3 + 5)^{1/2}$

2. $\dfrac{x}{\sqrt{x^2 + 1}}$

3. $g' \geq 0$ since $\cos x \geq -1$

4. $-\sqrt{3}$

5. $\dfrac{1 + \sqrt{3}}{2}$

6. $6\sin x(1 - 2\cos x)^2$

7. 10

8. $\frac{1}{4}$ units^2

SET 91

1. $2anx(ax^2 + b)^{n-1}$
2. $\dfrac{x}{\sqrt{(x+1)(x-1)}}$
3. $2x\cos(x^2 + 2)$
4. -1
5. $1 + \frac{\pi}{2}$
6. $-6\sin 2x \cos^2 2x$
7. (i) $-\cos \alpha + c$ (ii) $\frac{-1}{2}\cos(2\alpha + 1) + c$ (iii) $\frac{1}{3}\cos(2 - 3\alpha) + c$
8. 36 units^2

SET 92

1. $4x(3x^2 - 5)^{-1/3}$
2. $\dfrac{1-4x}{2(2x^2 - x - 6)^{3/2}}$
3. $-6x\sin(3x^2 - 5)$
4. $\frac{4}{25}$
5. 0
6. $-2\sin(2x + 3)$
7. $\frac{\pi}{2}$
8. $1\frac{1}{3}$ units^2

SET 93

1. $\log_e y = x$
2. $\frac{1}{2}$
3. $x = 8$
4. $\log_3 27t$
5. $2 + \log_{10} 1{\cdot}5$
6. 0·0182
7. $r = 38\, s^7$
8. $y = -\log_{10} x$

SET 94

1. $r = e^p$
2. $\frac{3}{2}$
3. 4
4. $\log_2 2k$
5. $1 + \log_6 4$
6. 0·0157
7. $t = 5u^2$
8. $y = \log_5 (-x)$

SET 95

1. $\log_2 y = x$
2. 3
3. 81
4. $\log_3 6$
5. $1 - \log_{10} 2$
6. 622 units
7. $p = 4q^4$
8. $y = \log_{10}(x - 1)$

SET 96

1. $2^p = q$
2. 1
3. 2·41
4. $\log \dfrac{x^3}{y^2}$
5. (0·1, 0)
6. 466 mg
7. $u = 2{\cdot}5\, v^{1{\cdot}5}$
8.

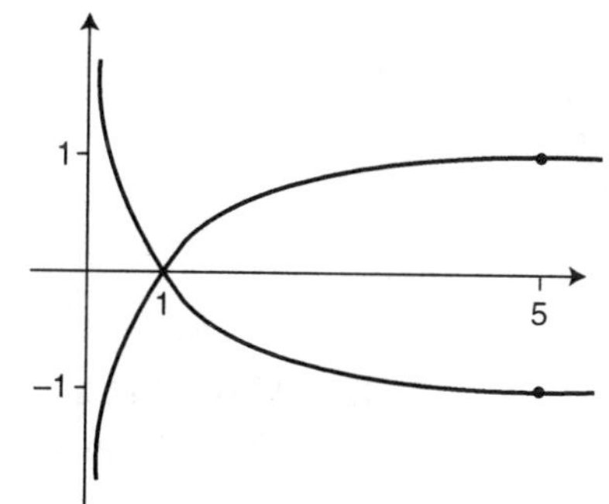

SET 97

1. $\log_q 3 = n$
2. $\frac{5}{2}$
3. 1·77
4. 0
5. $\frac{1}{2}\log 4 + 3\log 3$
6. 35 m
7. $y = 10x^6$
8. $y = \log_3 (2 - x)$

SET 98

1. $y = a^p$
2. 5
3. 32
4. $\log_e 5e^2$
5. $\left(\frac{1}{2}, 0\right)$
6. 99 years
7. $y = 3{\cdot}2 \times 10^{0{\cdot}3x} = 3{\cdot}2 \times 2^x$
8. 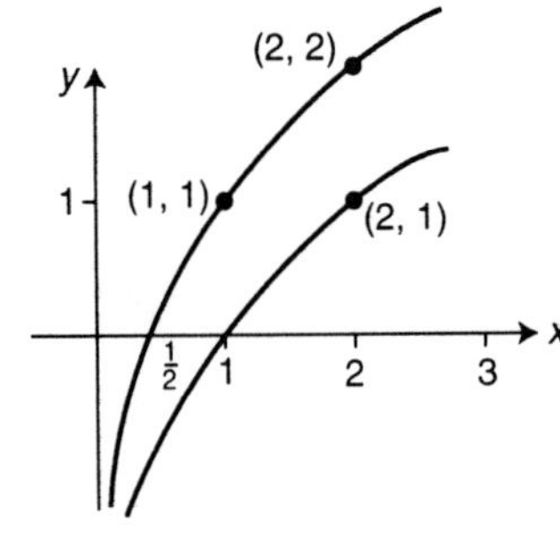

SET 99

1. 4, 30
2. –4, 210
3. $k = 5$; $\alpha = 53{\cdot}1$
4. $\sqrt{13}\cos(x - 56{\cdot}3)$
5. Max $\sqrt{13}$ at $x = 56{\cdot}3$; Min $-\sqrt{13}$ at $x = 236{\cdot}3$
6. 2, 30
7. 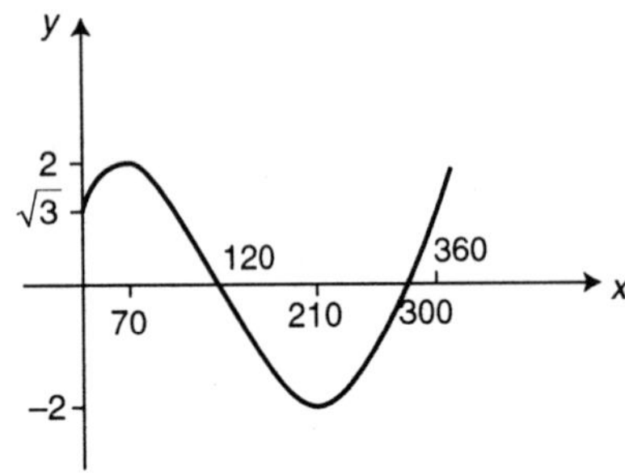

8. {41·6, 244·6}

SET 100

1. 7, 50
2. –7, 230
3. $k = 13$; $\alpha = 292{\cdot}6$
4. $3\sin(x + 48{\cdot}2)°$
5. Max 3 at $x = 41{\cdot}8$; Min –3 at $x = 221{\cdot}8$
6. $-\sqrt{13}$, 123·7
7. 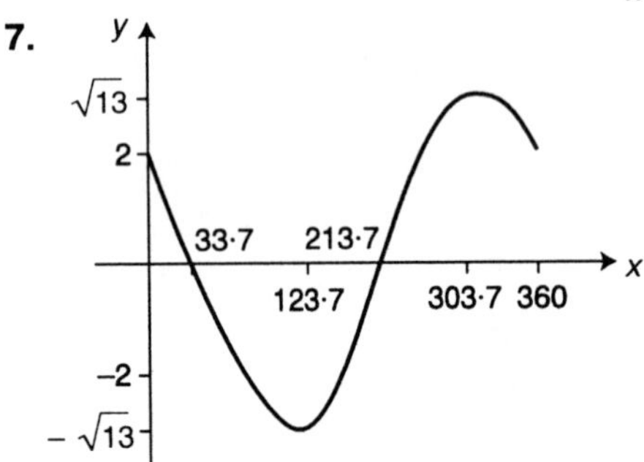

8. {82·5, 351·3}

SET 101

1. 5, 340
2. –1, 160
3. $k = \sqrt{2}$; $\alpha = 135$
4. $\sqrt{2}\sin(x + 45)°$
5. Max $\sqrt{2}$ at $x = 45$; Min $-\sqrt{2}$ at $x = 225$
6. –5, 233·1
7. 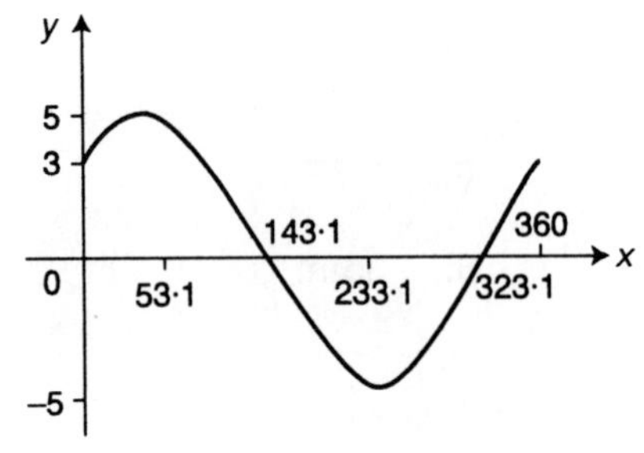

8. {130·2, 342·4}

SET 102

1. 6, 105
2. 4, 15
3. $k = 2$; $\alpha = 225$
4. $5\cos(x + 36{\cdot}9)°$
5. Max 5 at $x = 323{\cdot}1$; Min -5 at $x = 143{\cdot}1$
6. $-\sqrt{2}, \dfrac{3\pi}{4}$
7. 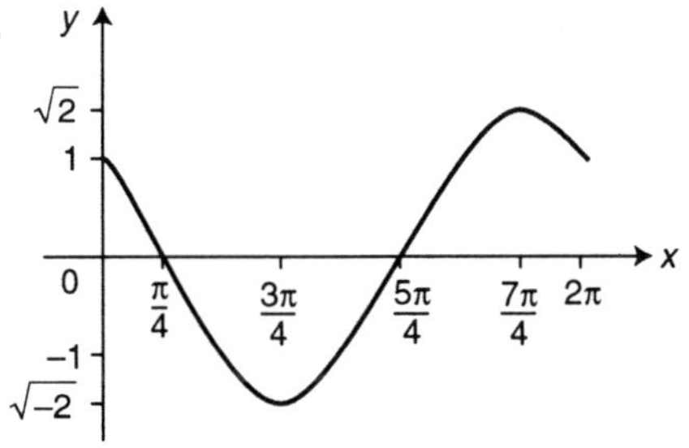

8. {0·971, 2·931}

SET 103

1. $3, \dfrac{\pi}{6}$
2. $-1, \dfrac{7\pi}{6}$
3. $k = \sqrt{2}$; $\alpha = 315$
4. $\sqrt{13}\cos(x - 33{\cdot}7)°$
5. Max $\sqrt{13}$ at $x = 48{\cdot}2$; Min $-\sqrt{13}$ at $x = 228{\cdot}2$
6. $2, \dfrac{\pi}{3}$
7. 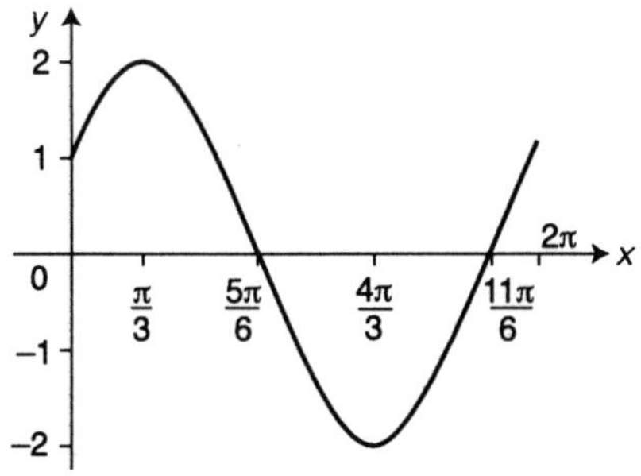

8. {257·6, 349·8}

SET 104

1. $5, \dfrac{5\pi}{4}$
2. $-1, \dfrac{\pi}{4}$
3. $k = \sqrt{13}$; $\alpha = 4{\cdot}12$
4. $\sqrt{34}\cos(x + 59)°$
5. Max $\sqrt{34}$ at $x = 301$; Min $-\sqrt{34}$ at $x = 121$
6. $\sqrt{2}, \dfrac{3\pi}{4}$
7. 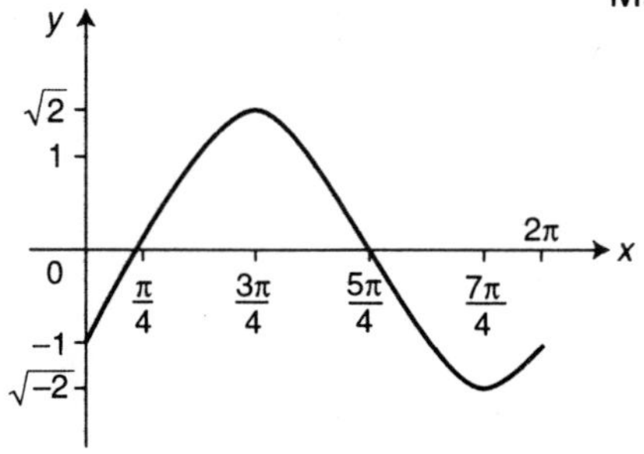

8. {39·2}

SET 105

1. -5
2. $\sqrt{5}\cos(x + 26{\cdot}6)°$
3. log 6
4. $5\frac{1}{3}$
5. e.g.,
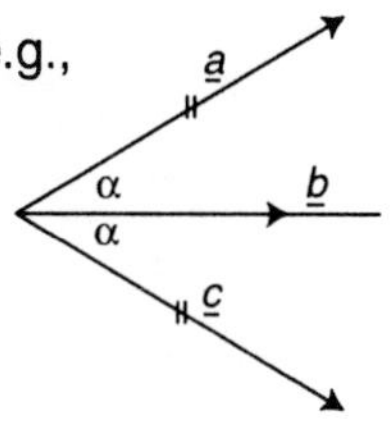

6. 2π
7. $b = 3$, $a = 9$
8. $\dfrac{1}{2(1 - x)^{3/2}}$

SET 106

1. Parallelogram
2. 5, 40
3. $\log_3 45$
4. $\dfrac{-1}{2(x-2)}$
5. Obtuse
6. 6
7. $a = 2,\ b = 5$
8. $4x\cos(2x^2 + 1)$

SET 107

1. (4, 1, 2)
2. $\sqrt{34}\cos(x - 59)°$
3. $\dfrac{1}{4}$
4. $\dfrac{1}{2}\sin(2x + 1) + c$
5. 2
6. –9
7. $p = 11q^3$
8. $-5\sin(5x + 4)$

SET 108

1. (8, –1, 5)
2. $3 - \sqrt{2}$, 160
3. $3\log 2 - 2\log 3$
4. $\dfrac{-2}{15(5x+3)^3} + c$
5. 9, 2, 3, 90
6. $\sqrt{13}\sin(x + 56{\cdot}3)°$
7. $\log_e s = 4t + \log_e 5$
8. $\dfrac{3}{2}(2x - 5)(x^2 - 5x - 2)^{1/2}$

SET 109

1. F **2.** F **3.** F **4.** F **5.** F **6.** F **7.** F **8.** T

SET 110

1. (i)

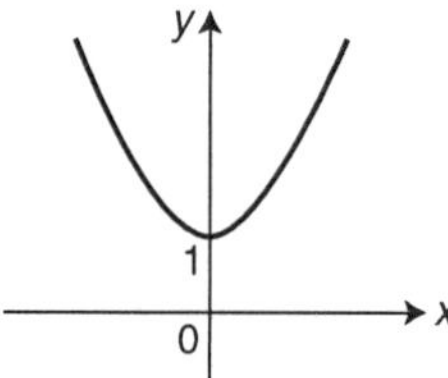

(ii)

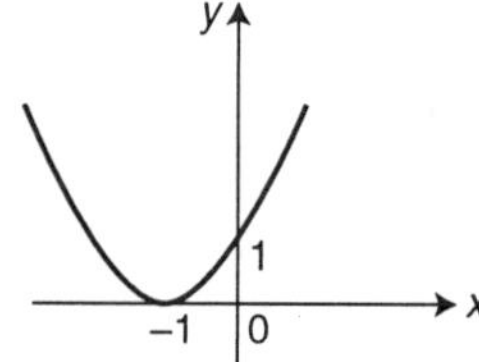

2. $x = h$
3. $f(p) = 0$
4. $h^2 + k^2 < r^2$
5. $a_1b_1 + a_2b_2 + a_3b_3$
6. (i) $\cos A$ (ii) $\sin A$
7. $a = 2,\ b = 1,\ c = 0$
8.

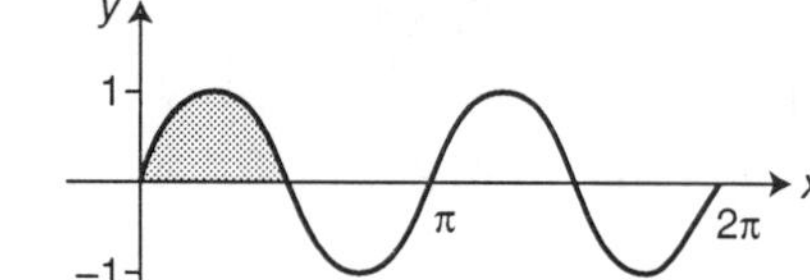

SET 111

1. (i) $\dfrac{1}{2x}$ (ii) $\dfrac{2}{x}$
2. (i) $\sin^2 A$ (ii) $\cos^2 A$
3. $b = a^x$
4. $x \neq k,\ x \neq l$
5. Parallelogram
6. $\sqrt{(x_2 - x_1)^2 + (y_2 - y_1)^2 + (z_2 - z_1)^2}$
7. (i) $\dfrac{1}{2}(1 + \cos 2A)$ (ii) $\dfrac{1}{2}(1 - \cos 2A)$
8. 2

SET 112

1. When $-1 < a < 1$
2. $q^2 - 4pr < 0$
3. $\sqrt{a^2 + b^2 + c^2}$
4. (i) $\left(\frac{r_1 + s_1}{2}, \frac{r_2 + s_2}{2}, \frac{r_3 + s_3}{2}\right)$ (ii) $r_1s_1 + r_2s_2 + r_3s_3$
5. $\frac{5}{4}$
6. (i) $a = 1$, $b = 2$, $c = 0$ (ii) π
7. $10\sin 5x \cos 5x$ ($= 5\sin 10x$)
8. $\frac{x}{2} - \frac{1}{20}\sin 10x + c$

SET 113

1. (i) 1 (ii) 1 (iii) 1
2. $\log_a b = x$
3. $\left(\frac{a}{2}, \frac{b}{2}\right)$
4. Collinear
5. (i) 1 (ii) $-\cos 2A$ (iii) $\sin 2\theta$
6.
7. $-\sin x + \cos x$
8. $k + l$

SET 114

1. (i)

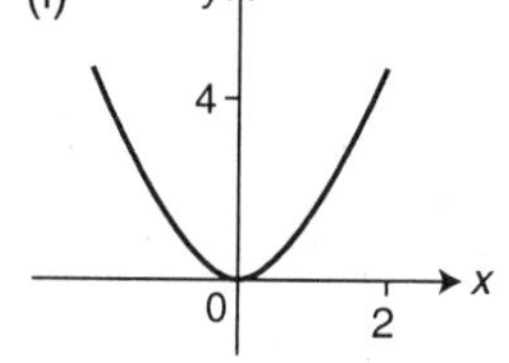

(ii)

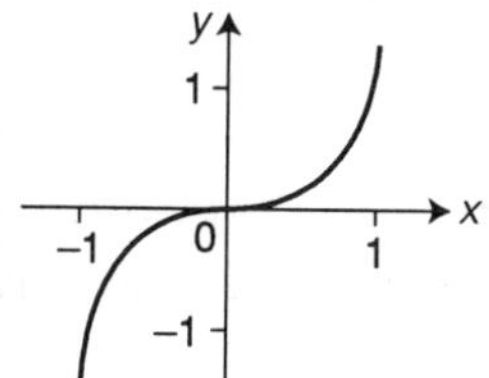

2. (i) 1 (ii) 1 (iii) 0
3. 1, 0
4. $\frac{y_2 - y_1}{x_2 - x_1}$
5. (i) $\cos (A + B)$ (ii) $\sin (Y - X)$
6. (i) 2 (ii) 3 (iii) 1
7. $f'(a)$
8. 0

SET 115

1. $x = \frac{-k \pm \sqrt{k^2 - 4hl}}{2h}$
2. $\left(\frac{x_a + x_b}{2}, \frac{y_a + y_b}{2}\right)$
3. (i) $a^{-n} = \frac{1}{a^n}$ (ii) $a^{1/n} = \sqrt[n]{a}$ (iii) $\log_a X - \log_a Y = \log_a \frac{X}{Y}$ (iv) $\log_b XY = \log_b X + \log_b Y$
4. $y - k = n(x - h)$
5. $\mathbf{p} = \frac{1}{3}(\mathbf{m} + 2\mathbf{n})$
6. (i) π (ii) $\frac{\pi}{2}$ (iii) $\frac{2\pi}{3}$ (iv) 4π (v) 8π
7. (i) $\cos x$ (ii) $-\cos x + c$
8. $\int_a^b g(x)\, dx$

SET 116

1. $x = 1$ **2.** $\log y = n \log x + \log a$ **3.** (i) $\frac{\sqrt{a}}{a}$ (ii) $\frac{\sqrt{a} + b}{a - b^2}$

4. $\frac{-1}{m}$

5.

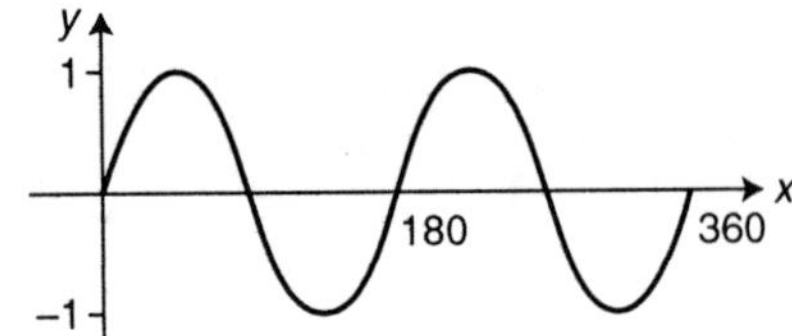

6. 5

7. $f'(x) > 0$

8. $-\frac{1}{2}\cos 2x + c$

SET 117

1. $\frac{5}{3}$ **2.** $ah^2 + bh + c$ **3.** $c = 0$

4. 0 **5.** (i) $2\cos^2 A - 1$ (ii) $2\cos^2 2A - 1$ (iii) $1 - 2\sin^2 \frac{A}{2}$

6. (i) 1 (ii) 2 (iii) 2 **7.** $f'(b)$ **8.** $2k$

SET 118

1. (0, 1) **2.** $\frac{1}{2}qr \sin P$ **3.** $\log_a XYZ$

4. $(x - u)^2 + (y - v)^2 = r^2$ **5.** (i) 1 (ii) $\cos x$ (iii) $\sin x$ **6.** (i) 2π (ii) π (iii) $\frac{2\pi}{3}$

7. $na(ax + b)^{n-1}$ **8.** $\int_a^c f(x)\, dx$

SET 119

1. $x^3 - 6x^2 + 8x + 4$ **2.** $f(x) = kx^2(x - 4)$; $k = -1 \Rightarrow f(x) = 4x^2 - x^3$

3. (i) a line from A to the mid-point of BC (ii) a line from B, perpendicular to AC

4. (i) 2 (ii) 6 **5.** 1 **6.** (i) $\frac{1}{2}$ (ii) -1

7. $f'(a) = 0$ **8.** (i) $-\frac{1}{2}\cos 2x + c$ (ii) $\frac{1}{2}\sin 2x + c$

SET 120

1. (i) 2 (ii) 1 **2.** $y = mx + 4$ **3.** $\Delta \geq 0$

4. (i) $\log_a \frac{1}{n} = -\log_a n$ (ii) $\log_a X^n = n \log_a X$ **5.** $(-h, -k)$; $\sqrt{h^2 + k^2 - l}$

6. $|\mathbf{u}| = 0$, $|\mathbf{v}| = 0$, $\mathbf{u} \perp \mathbf{v}$ **7.** Yes, it has length 1.

8. (i) $\frac{(ax + b)^{n+1}}{a(n + 1)} + c$ (ii) $-\frac{1}{a}\cos(ax + b) + c$

Printed by Bell & Bain Ltd., Glasgow, Scotland.